71
Electrical &
Electronic
Projects

For Beginners, Intermediate, and Engineering Students

Nikhil Shukla

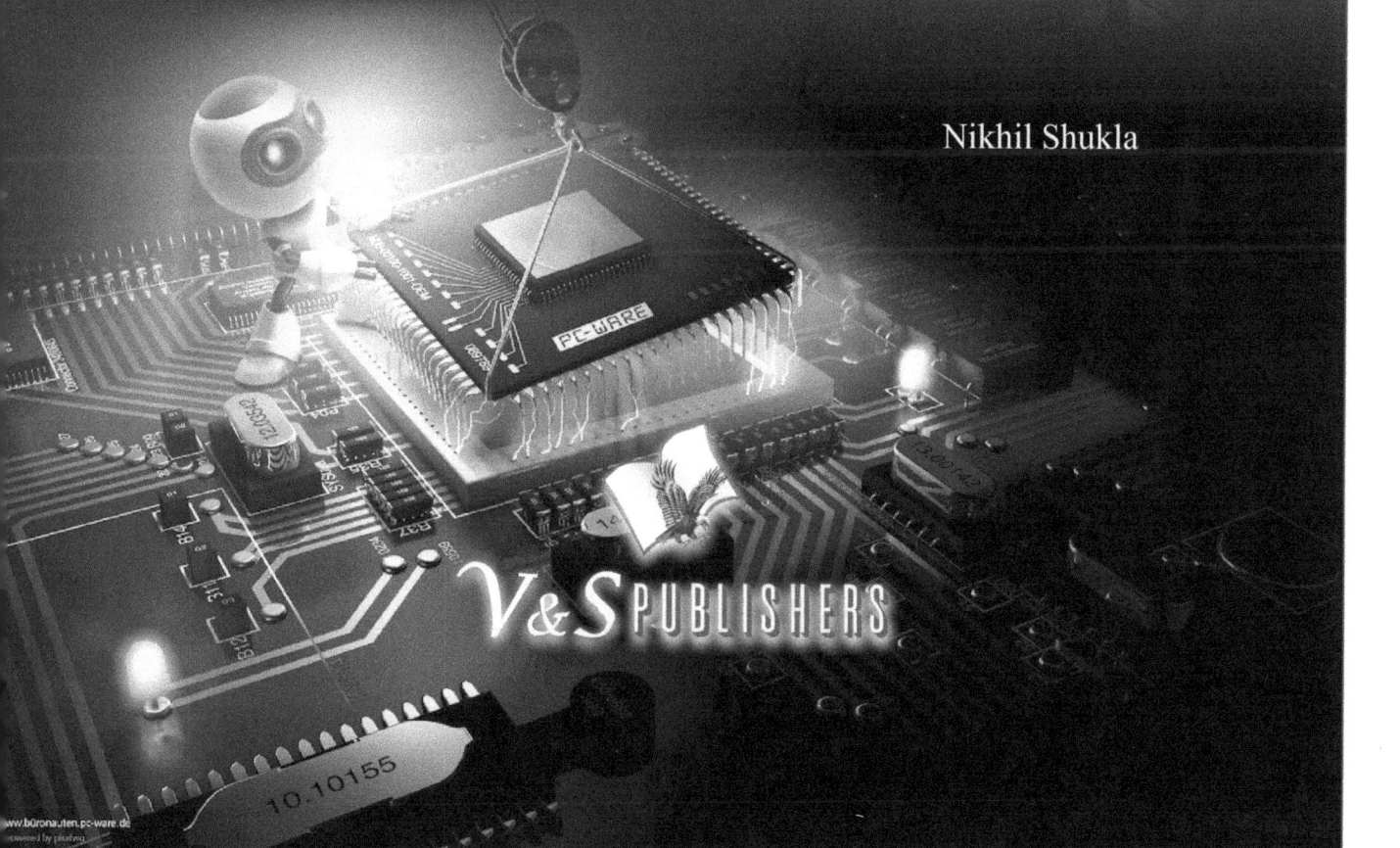

V&S PUBLISHERS

Published by:

V&S PUBLISHERS

F-2/16, Ansari road, Daryaganj, New Delhi-110002
23240026, 23240027 • *Fax:* 011-23240028
Email: info@vspublishers.com • *Website:* www.vspublishers.com

Regional Offi ce : Hyderabad
5-1-707/1, Brij Bhawan (Beside Central Bank of India Lane)
Bank Street, Koti, Hyderabad - 500 095
040-24737290
E-mail: vspublishershyd@gmail.com

Branch Offi ce : Mumbai
Godown # 34 at The Model Co-Operative Housing, Society Ltd.,
"Sahakar Niwas", Ground Floor, Next to Sobo Central, Mumbai - 400 034
022-23510736
E-mail vspublishersmum@gmail.com

Follow us on: t f in

All books available at **www.vspublishers.com**

Printed at : Param Offseters, Okhla, New Delhi-110020

Chapter - 2

Electrical and Electronic Components and Their Symbols

There are a large number of symbols used to represent a large range of electronic and electrical components. It is essential that you recognise the common components and learn what they actually do. Various components with their symbols are given below:

Component	Image	Symbol
Slide Switch		
Micro Switch		
Toggle Switch		
Dry Reed Switch		
Push Switch		
Relay		RL RELAY
Thyristor		

Light Emitting Diode		
Diode		
Bulb/Filament		
Motor		M
Solenoid		
Variable Resistor		
Preset Resistor		

LDR / Light Dependent Resistor		
Resistor		
Thermistor		
Micro Phone		
Buzzer		
Speaker		

Symbol	Component Name	Meaning/Uses
	Jumper	
		pins
		Solder to close connection
Ground Symbols		
		electrical shock protection
Resistor Symbols		
	Potentiometer	
Capacitor Symbols		
		Electrolytic capacitor
Inductor / Coil Symbols		
	Inductor	

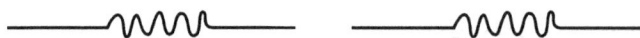

	Variable Inductor	
Power Supply Symbols		
	Voltage Source	
		mechanical rotation of the generator
	Source	current of other circuit element
	Source	current of other circuit element
Meter Symbols		
	Voltmeter	
		Measures resistance
		Measures electric power
Lamp / Light Bulb Symbols		
Diode / LED Symbols		
		drop

		Variable capacitance diode
Transistor Symbols		
	Transistor	at base (middle)
	tor	base (middle)
		gain of the product of each gain.

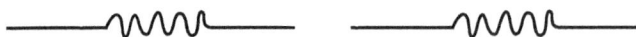

Logic Gates Symbols		
		Stores one bit of data
Miscellaneous Symbols		
	Motor	Electric motor
	Transformer	to high.
	Electric bell	

		The fuse disconnects when the current is from high currents.
		address.
	isolator	board
	Microphone	
	Schmitt Trigger	
		signal

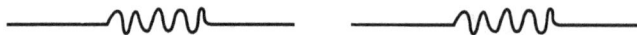

Chapter - 3

Choosing, Using and Testing
Electronic and Electrical Components

An electronic or electrical circuit is made up of a number of electronic components. Broadly, there are only five electronic components : three passive components and two active components as follows:

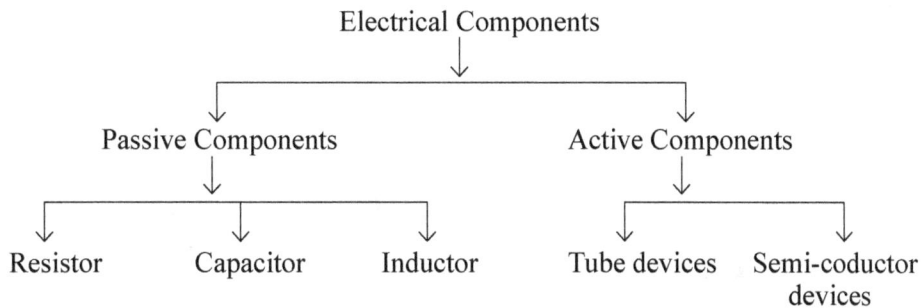

Electrical Components

Passive Components   Active Components

Resistor   Capacitor   Inductor   Tube devices   Semi-coductor devices

These are the backbone of any electronic construction project. Before we start building projects and do some construction activitiy, a brief introduction to these components, knowledge of lead indentification, colour coding schemes, their functions are essential startup.

Passive Components

Passive components are the components which are not capable of processing or amplifying an electrical signal but these components are an important part of a circuit because without them an active component is not able to amplify or process a signal. These components couduct current in both directions; and hence, they are known as bidirectional devices. Passive components are of three types :

(i) Resistors
(ii) Capacitors
(iii) Inductors

Resistors

Linear Resistors

       (ii) Variable resistors

Fixed resistors:
with screws.

resistors.

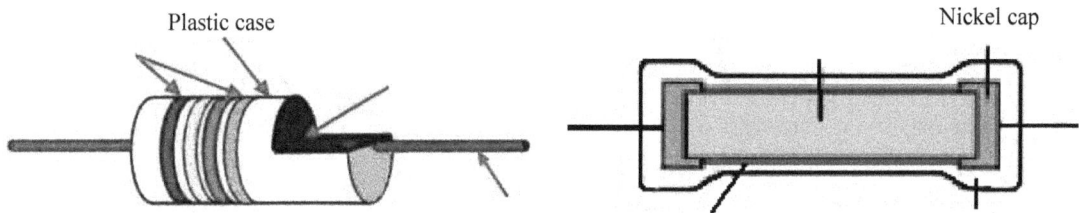

Plastic case

Nickel cap

Helical cut to reach the

End caps
with leads

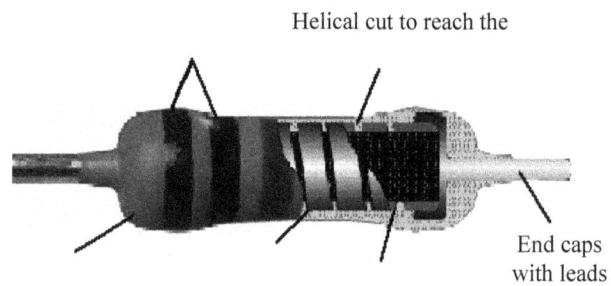

more than one layer on an insulating core. Typical resistance range for wire wound resistors is

Variable linear resistors:

    (a) Variable wire resistor
    (b) Potentiometer
    (c) Trimmer

(a) Varible wire-wound resistors: These are manufactured from nichrome wire wound on

ranging from 1

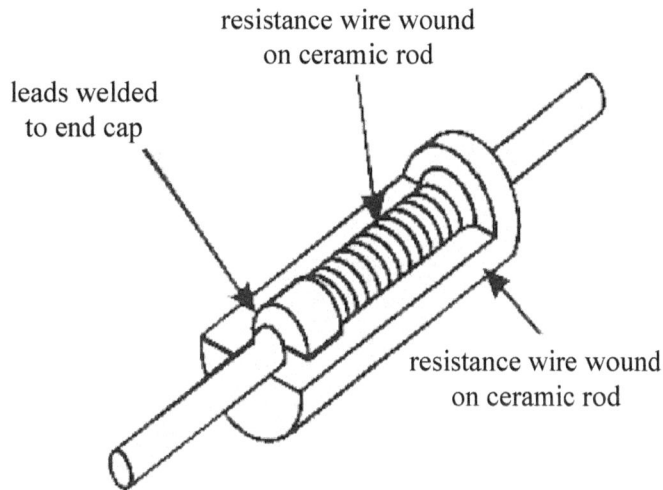

leads welded to end cap

resistance wire wound on ceramic rod

resistance wire wound on ceramic rod

(b) Potentiometer:

or plastic enclosure. In some cases, the tolerance is also indicated.

Single-turn trimmer multi-turn trimmer

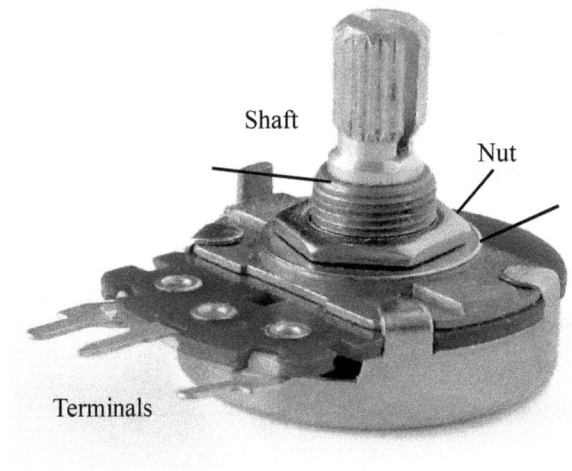

Shaft
Nut
Terminals

Trimmers: Trimmers are used in electronic circuit to trim the circuit to the desired operating

Single-turn Trimmer

Multi-turn Trimmer

Non-linear Resistors

Non-linear resistors are manufactured from semi-conductor materials. The non-linear quality of

    (i)   Thermistors
    (ii) Photoresistors
    (iii) Varistors

(i) **Thermistors:**

widely used.

the resistance decreases with the increase in temperature. Thermistors are made in the form of brobes, beads, disc, washers and rods. The beads are made in diameter

power applications.

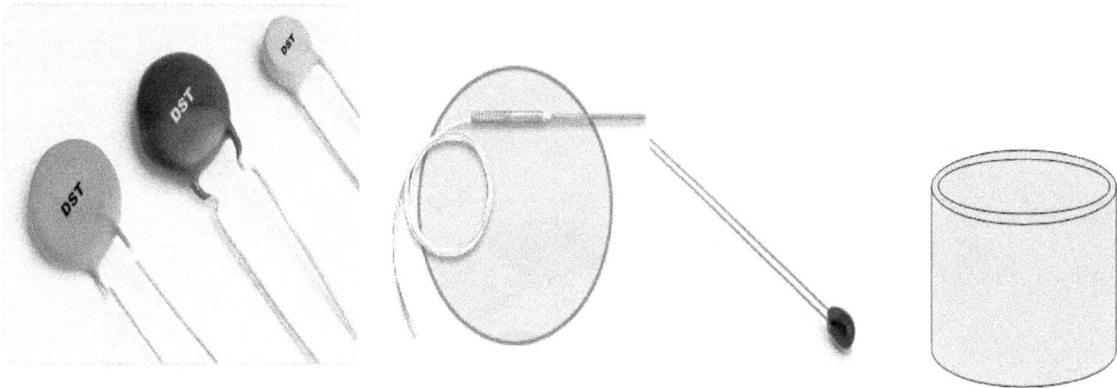

Probe Thermistor

(ii) **Photoresistors:**
are manufactured from semi-conductor materials whose resistance changes when

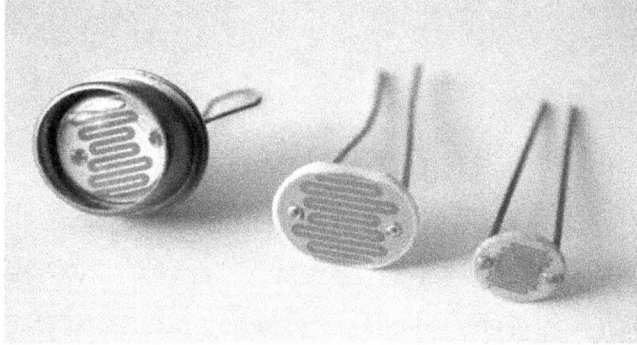

(iii) **Varistors:**

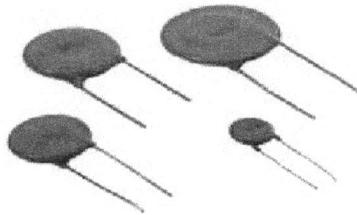

Resistor designation :

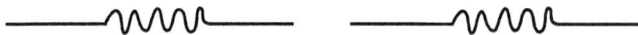

Colour Code Chart for Fixed Composition and Film Resistors — MIL and EIA Standard Values*

Standard Colour Code

First Band 1st Digit	
Colour	Digit
Black	0
Brown	1
Red	2
Orange	3
Yellow	4
Green	5
Blue	6
Violet	7
Gray	8
White	9

Second Band 2nd Digit	
Colour	Digit
Black	0
Brown	1
Red	2
Orange	3
Yellow	4
Green	5
Blue	6
Violet	7
Gray	8
White	9

Third Band Multiplier	
Colour	Multiplier
Black	1
Brown	10
Red	100
Orange	1,000
Yellow	10,000
Green	100,000
Blue	1,000,000
Silver	0.01
Gold	0.1

Fourth Band Resistance Tolerance	
Colour	Tolerance
Silver	± 10%
Gold	± 5%
*No Band	± 20%

Fifth Band Reliability Level (Percent per 1,000 Hours)	
Colour	Level
Brown	M = 1.0%
Red	P = 0.1%
Orange	R = 0.01%
Yellow	S = 0.001%
*Mil-R-39008 Resistors Only	

Capacitors

etc.

Fixed capacitors

constant. These capacitors are most widely used in radio and telecommunication applications.

Electrolytic capacitors

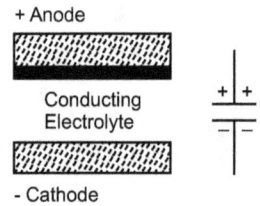

+ Anode

Conducting
Electrolyte

- Cathode

$+$ $+$

$-$

capacitor

capacitor is generally packed in metal cylinder. The cathode is connected to

(i) **Aluminium electrolytic capacitors :**

which acts as dielectric, formed on the anode foil surface.

(ii) **Tantalum electolytic capacitors :** Tantalum capacitors are far better than aluminium

low leakage of current.

Variable capacitors

Inductors

chokes for audio-frequency and radio-frequency ranges. These types of chokes are known as

Types of inductors

(ii) Variable inductors

(i) **Finxed inductors:**

   (a) air-core inductors

   (b) iron-core inductors

   (c) ferrite-core inductors

   (a) **Air-core inductors:** These are made of coils of wire-wound on a former made of simple cardboard. These

   (b) **Iron-core inductors:** Iron-core inductors are made of coils of wire wound on solid iron core. This iron-core is liminated to

   (c) **Ferrite-core inductors:**
   core inductors are made of coils of wire wound on a ferrite core. The

   Iron - core
   Inductors

   ferrite-core inductors are used at higher frequencies.

(ii) **Variable inductors :**

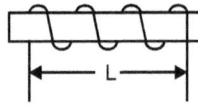

Inductance of a coil

Inductance of a coil can be determined with the help of its physical properties. The inductance of a coil

$$\frac{\mu_o \mu_r AN}{l}$$

where,

l

-7

r

Active Components

(i) Tube type
(ii) Semi-conductor or soild state tube type

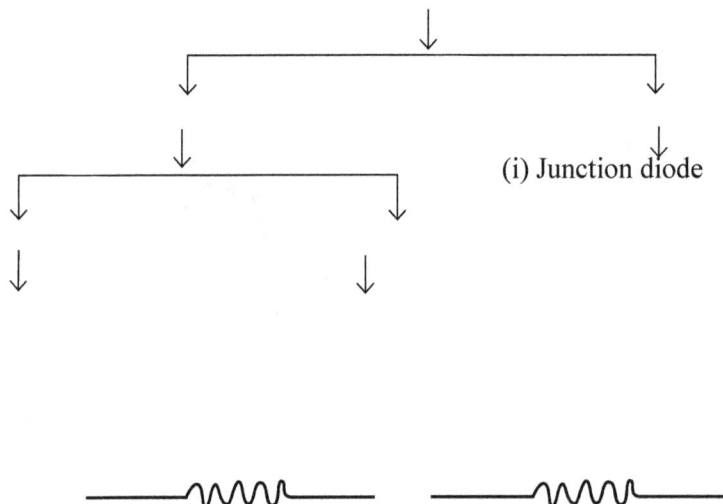

(i) Junction diode

Diodes :

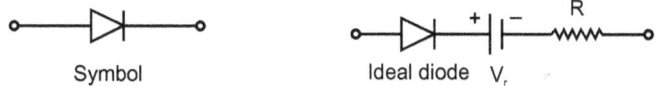

Symbol   Ideal diode V_r

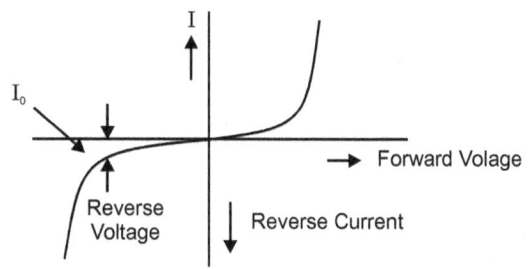

(ii) Zener diodes

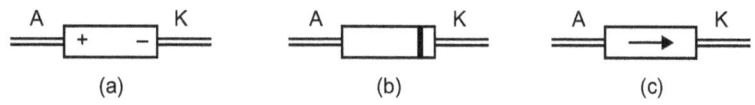

(a)   (b)   (c)

popular style of marking the leads of diodes is to put an arrow along the length of the diode with

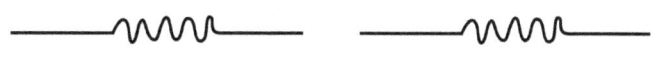

In case of Zener diodes made in these package styles, a band is put near cathodes. In some Zener

to the anode.

Testing Diodes

one with the lower resistance ranges. The multimeter leads are connected to the diode in such a way

diode is okay in the forward biased mode. Now, interchange the multimeter leads in an attempt to

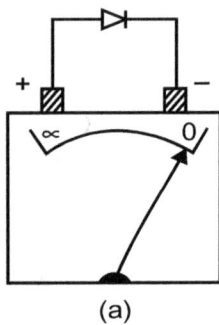

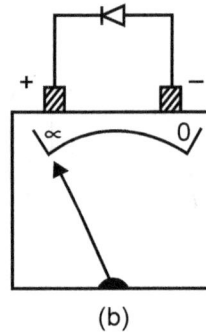

(a)

(b)

supply is gradually increased while monitoring

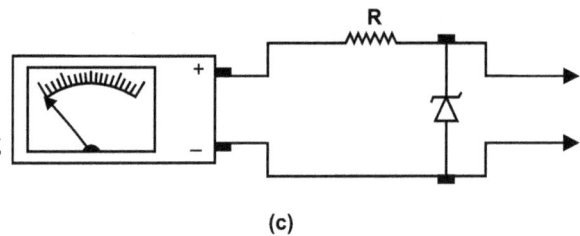

multimeter.

(c)

forces the Zener diode under test dissipates more power than it can safely handle. Typically, the current

Light Emitting Diodes

with one anode emits light of one colour and the same cathode with other anode emits light of the other colour.

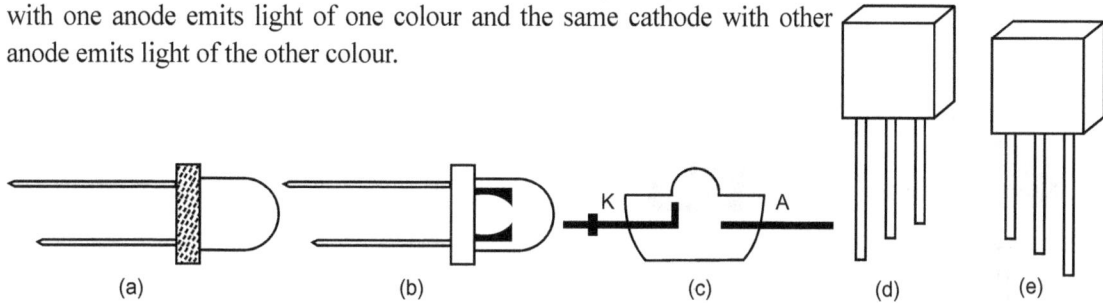

(a)   (b)   (c)   (d)   (e)

Testing the LED

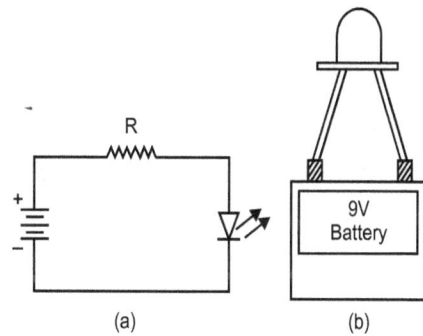

(a)   (b)

Tronsistors

Symbol

(a)   (b)

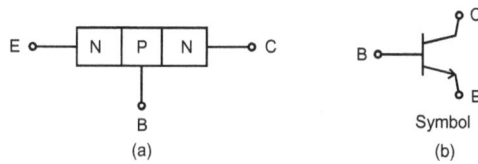

The another type is PNP in which a thin N-type semiconductor layer is sandwitched between two

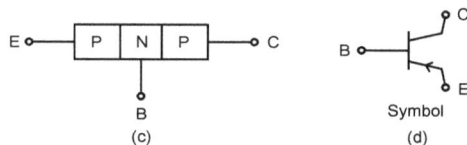

Symbol

(c)   (d)

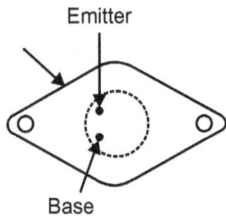

Emitter

Base

(TO-3)

(a)

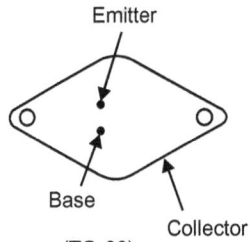

Emitter

Base

Collector

(TO-66)

(b)

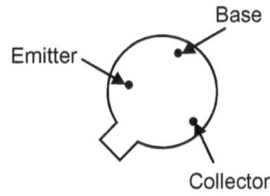

Emitter

Base

Collector

(TO-5, TO-18, TO-39)

(c)

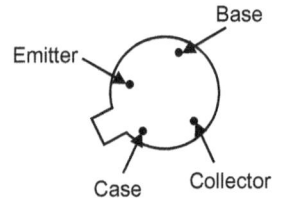

Emitter

Base

Case

Collector

(TO-72)

(d)

(TO-105, TO-106)

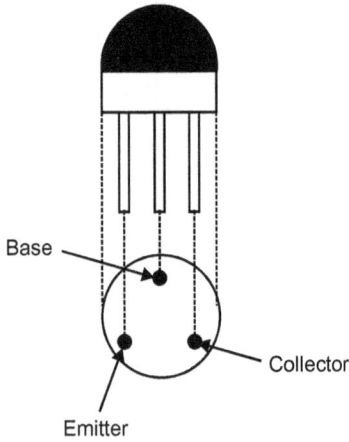

Base

Collector

Emitter

(e)

(TO-106)

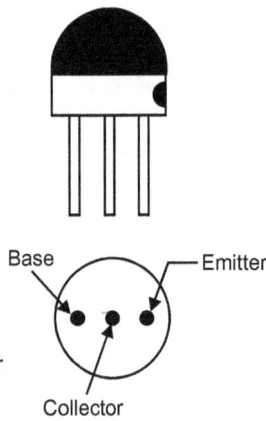

Base

Emitter

Collector

(f)

(TO-92)

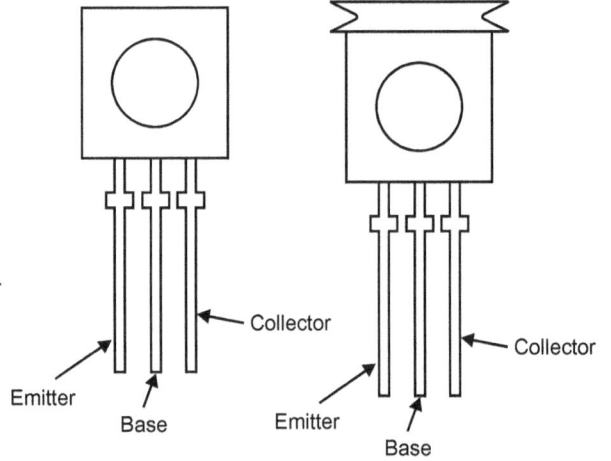

Collector

Emitter

Base

(g)

(TO-237)

Collector

Emitter

Base

(h)

(TO-220)

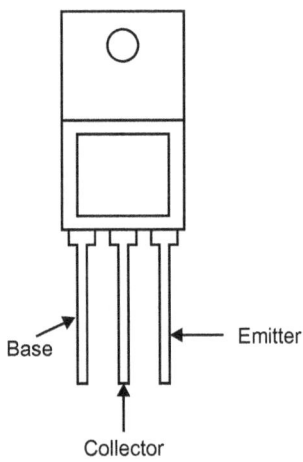

Base

Emitter

Collector

(i)

(TO-202)

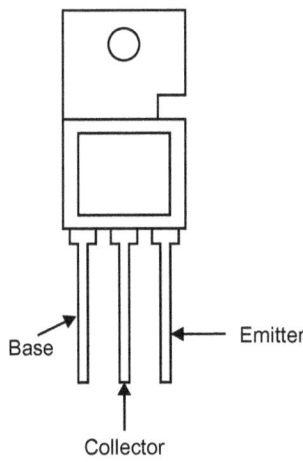

Base

Emitter

Collector

(j)

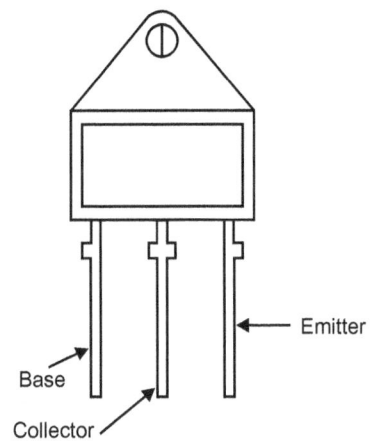

Base

Emitter

Collector

(k)

Lead Identification of Transistor Using Multimeter

The use of multimeter for indentifying the leads of bipolar transistors is similar to the one used for

shows an open circuit between emitter and collector leads in both the directions.

by making base-emitter function diode check.

NPN transistor.

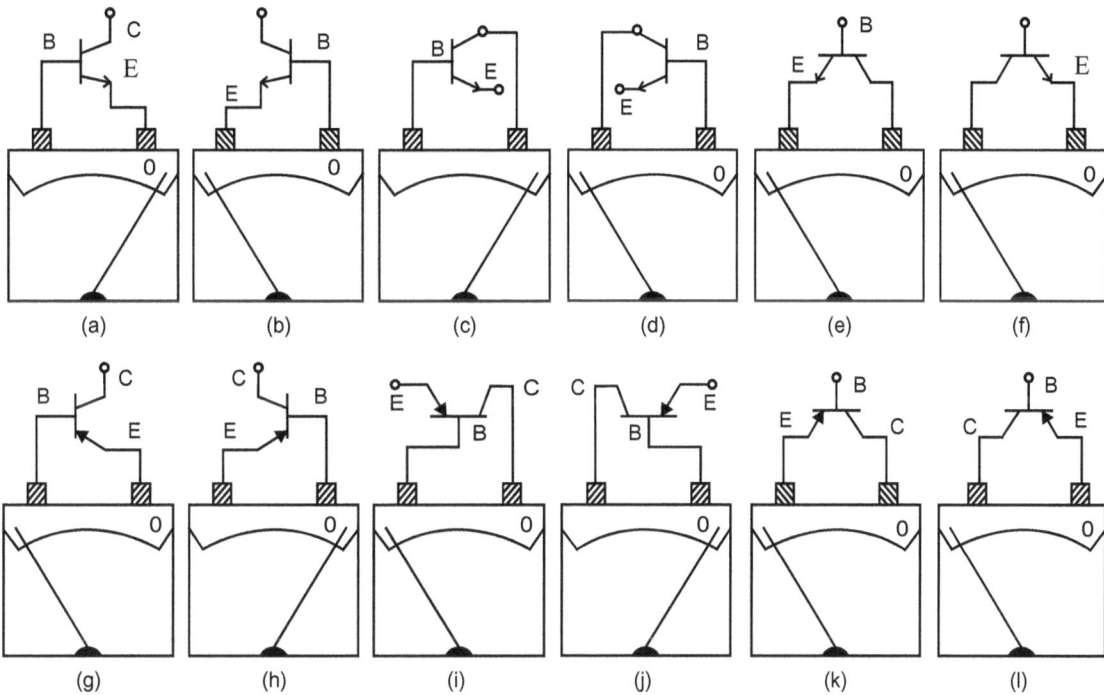

(a)	(b)	(c)	(d)	(e)	(f)

(g)	(h)	(i)	(j)	(k)	(l)

Testing Transistor

terminals should show an open condition in both directions.

Mechnical and Electromegnetic Devices
Mechnical Switches : Toggle Type

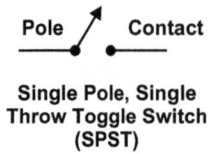

**Single Pole, Single
Throw Toggle Switch
(SPST)**

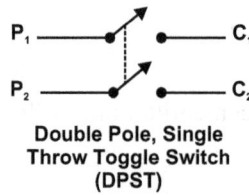

**Double Pole, Single
Throw Toggle Switch
(DPST)**

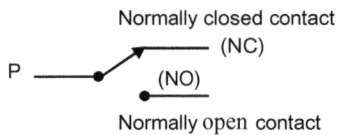

Throw Toggle Switch
(

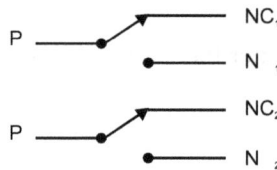

Throw Toggle Switch
(

what they are rated for, permanent damage can result to the switches.

used.

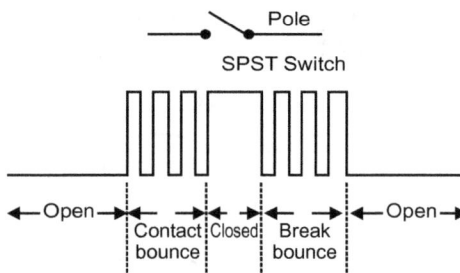

preferred.

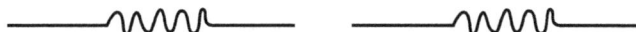

in physical structure and mode of operation between the toggle switches and micro-switches.

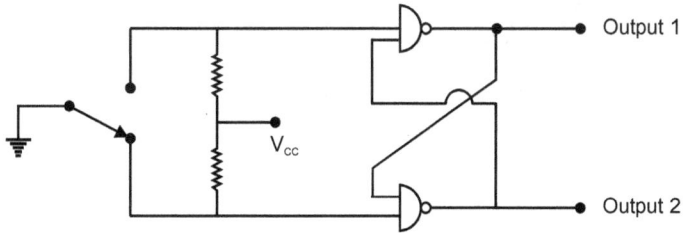

Electromagnetic Relays

Relay are most usefully employed when a number of independent circuits are switched on from

current is all that is needed to be supplied from digital logic, an important aspect in consideration of loading factors.

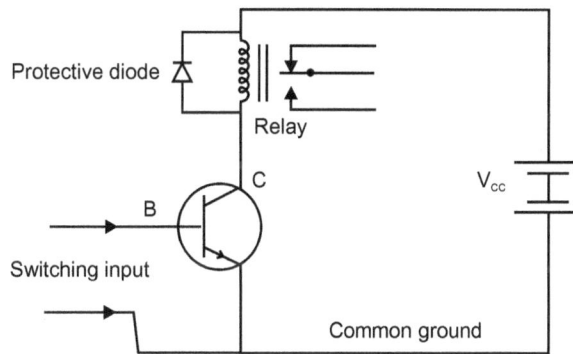

transistor.

The clamping diode offers a low impedance path to this spike and protects the transistor.

Reed Relays

Permanent Magnet

C_1 ——

C_2

Glass enclosure   Reeds

Electromagnet

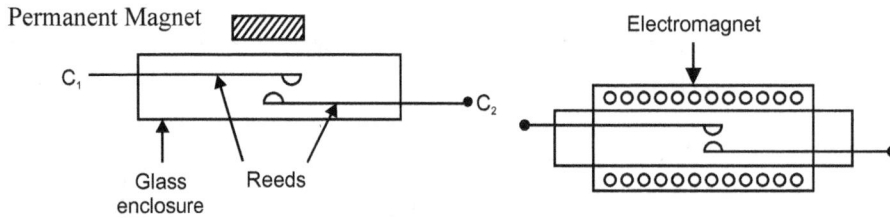

contact resistance.

contacts close and the relay circuit is completed.

action is required through electrical pulses, the reed relay is actuated by an electromagnet.

In communication systems, post acting reed relay are employed quite commonly. The inert atmosphere and the mercury witted contacts ensure that contact bounce, common with toggle switches, is not encountered.

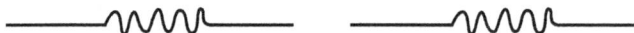

Chapter-4

PCB Fabrication and Design

Printed Circuit Board (PCB)

A printed circuit board popularly known as PCB is the first thing you would require when you device to build an electronic circuit.

A proper PCB ensures that various circuit components are interconnected as per the circuit diagram once they have been placed on the PCB in their proper positions and subsequently soldered. PCB design and fabrication techniques have been so much developed that it has become a subject in itself.

The projects or articles on construction given in various books on the subject and electronics magazine are usually complete in relevant technical or fabrication information including the PCB layout, the components layout, list of component parts and so on. What you need the most in such cases is only a familiarisation with the etching process. But in case you want to build a circuit for which a layout is not available, you will have to design the layout too.

PCB Fabrication

Different steps involved in the fabrication of a PCB are as follows :

1. Designing of Components Layout
2. Layout Designing of PCB
3. Transferring the PCB layout design onto the PC board laminate
4. Developing or etching the PCB
5. Operations like drilling, cutting, tinning etc.

Layout Designing of Components

The layout designing of components is the exercise of placement of different components constituting the circuit and then showing their interconnections as per the circuit diagram. The designing usually begins with an estimate of the size of the PCB (length and breadth) needed to accommodate various circuit components. The estimated dimensions are marked on graph paper and an attempt is made

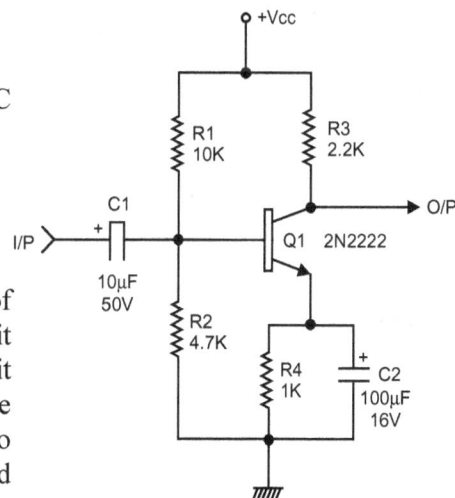

Fig : 4.1 Layout Designing of Components

to optimally place the components. It may be mentioned here that

components will largely depend upon the nature of interconnections and also on the circuit input, output, supply and ground points.

erent components making up the circuit is another pre-requistie to designing the components layout.

the interconnections can be made drawing lines (known as tracks). Ensure that the tracks are of optimum length and are not unnecessarily

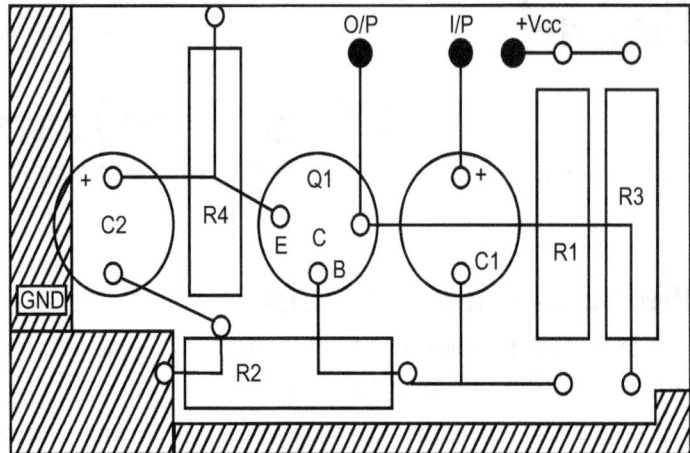

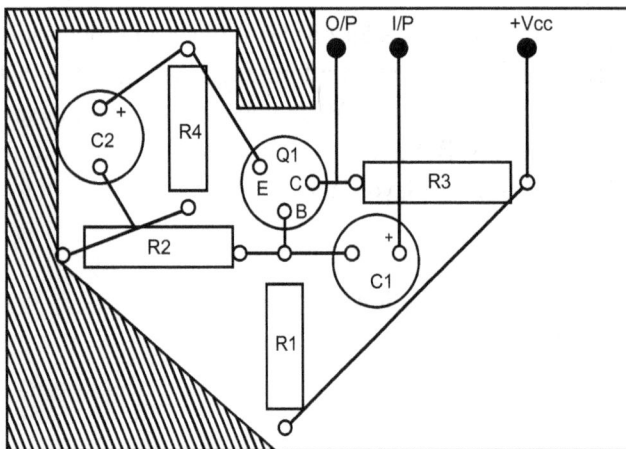

long. In fact, the components layout can be best drawn by placing components and making the interconnections side by

the components layout for the circuit

how an optimum components layout

design of components for the circuit of

increase in component count and the number of input and output connections.

practice does.

PCB-Layout Designing

the copper side. It can best be obtained by taking a carbon copy of the components layout already

Transferring PCB Layout Design onto the PCB Board Laminate

epoxy boards. Phenolic boards are much cheaper and are good enough for most of the commercial applications.

The next step is to thoroughly clean the copper side of the laminate with petrol or acetone or alcoholic spirit to make it completely free from any contaminants. Some cleaning sprays are also commercially

you need to do is to only place the dry transfer at the desired place and rub it against copper. The

pattern has been drawn with the paint, it should be allowed to fully dry up before it is etched. (See

Developing or Etching the PCB

chloride solution. The solution is prepared by

The solution should be nicely stirred and a few drops of hydrochloric acid be added if you want to speed up the etching process. The tray can be

solution to repeatedly interact with the copper side of the laminate. The etching process may take about half an hour to one hour, the etching

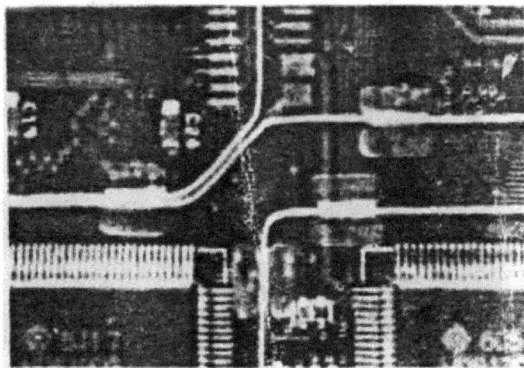

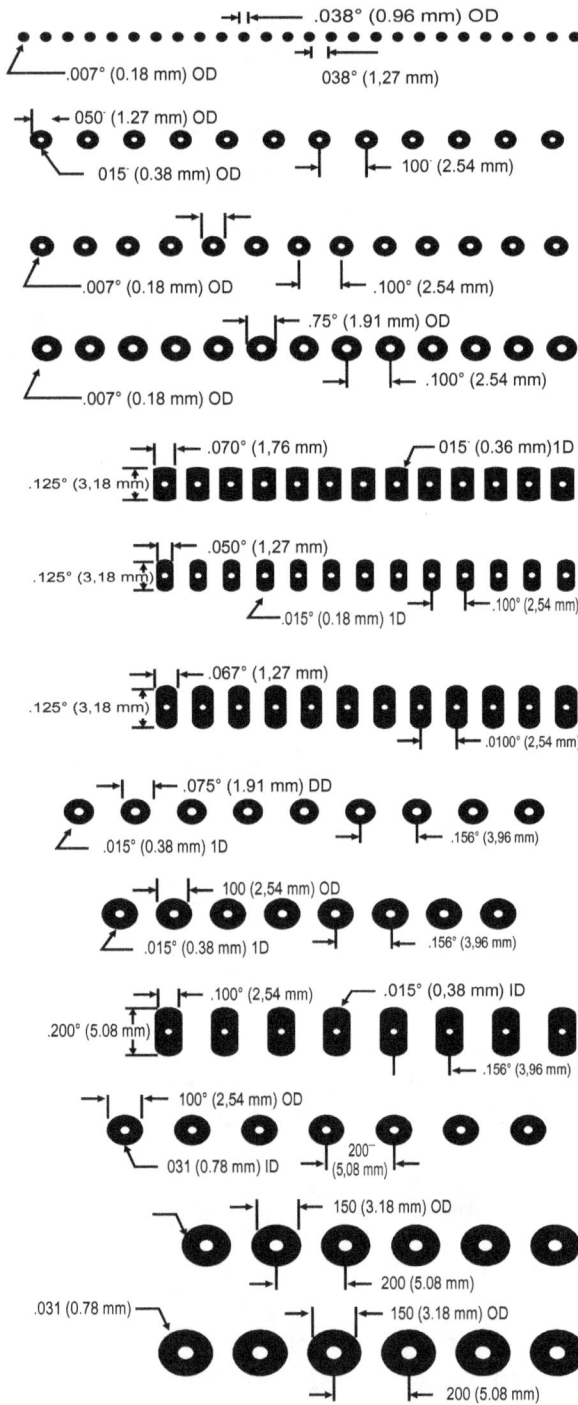

.038° (0.96 mm) OD

.007° (0.18 mm) OD   038° (1,27 mm)

050' (1.27 mm) OD

015' (0.38 mm) OD   100' (2.54 mm)

.007° (0.18 mm) OD   .100° (2.54 mm)

.75° (1.91 mm) OD

.007° (0.18 mm) OD   .100° (2.54 mm)

.070° (1,76 mm)   015' (0.36 mm)1D
.125° (3,18 mm)

.050° (1,27 mm)
.125° (3,18 mm)   .015° (0.18 mm) 1D   .100° (2,54 mm)

.067° (1,27 mm)
.125° (3,18 mm)   .0100° (2,54 mm)

.075° (1.91 mm) DD
.015° (0.38 mm) 1D   .156° (3,96 mm)

100 (2,54 mm) OD
.015° (0.38 mm) 1D   .156° (3,96 mm)

.100° (2,54 mm)   .015° (0,38 mm) ID
.200° (5.08 mm)   .156° (3,96 mm)

100° (2,54 mm) OD
031 (0.78 mm) ID   200" (5,08 mm)

150 (3.18 mm) OD
200 (5.08 mm)

.031 (0.78 mm)   150 (3.18 mm) OD
200 (5.08 mm)

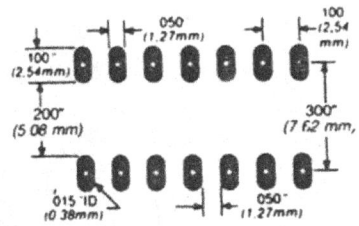

050 (1.27mm)   100 (2.54 mm)
100" (2.54mm)   200° (5 08 mm)   300° (7 62 mm,)
015" ID (0 38mm)   050' (1.27mm)

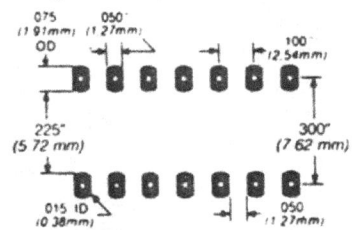

075 (1 91mm) OD   050' (1.27mm)   100' (2.54mm)
225° (5 72 mm)   300° (7 62 mm)
015 ID (0 38mm)   050 (1 27mm)

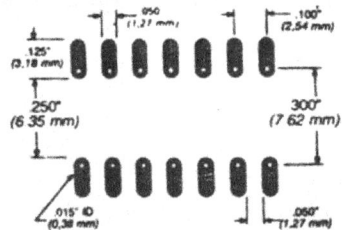

050 (1.27 mm)   .100° (2,54 mm)
.125° (3.18 mm)
250° (6 35 mm)   300° (7 62 mm)
015° ID (0,38 mm)   050° (1.27 mm)

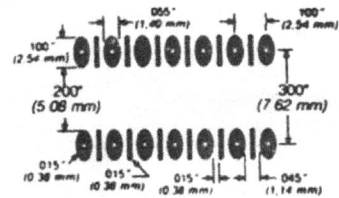

055' (1.40 mm)   100" (2.54 mm)
100" (2 54 mm)   200° (5 08 mm)   300° (7 62 mm)
015" (0 38 mm)   015' (0 38 mm)   015' (0 38 mm)   045° (1.14 mm)

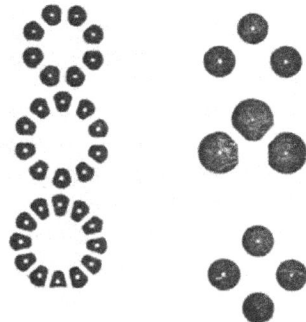

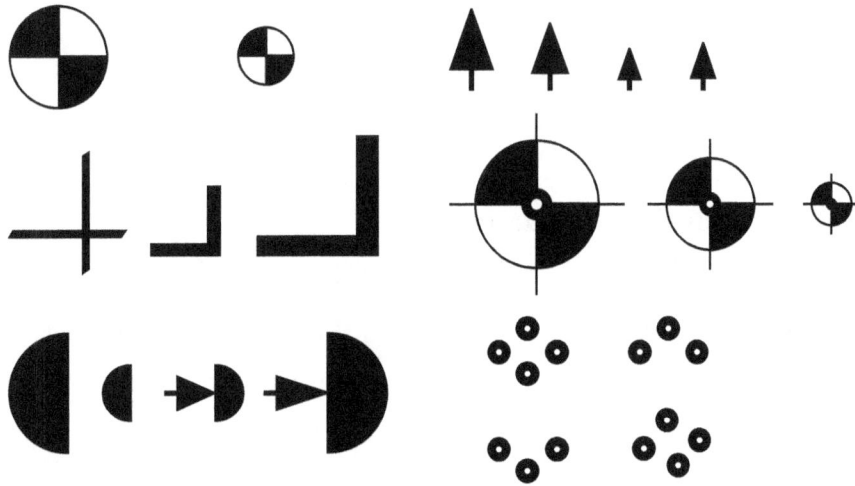

the etching is complete, the solution is likely to penetrate the copper portions that you wanted to

Solution once made will certainly last for three to four etchings.

The laminate is throughly washed in water after

with alcohol or acetone.

Other Operations Like Drilling, Cutting, Tinning etc. :

from component to component. It is 1 mm for

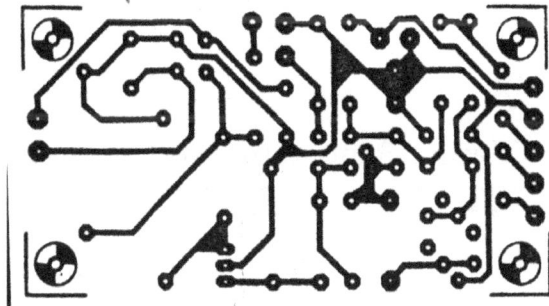

Soldering Techniques

Some Useful Tips for Good Soldering

1. The two surfaces to be soldered should be thoroughly cleaned and made free from any dust, and proper tinning of the components are essential for good soldering results.

between the solder and the metal being soldered. If hot solder is applied to a cold metal or a

will be proper only when the alloy to be soldered is hot enough to remain in a liquid state

terminal end slightly remember not to exceed the component lead temperature beyond

and instead of applying it on the soldering iron tip. Now with the soldering iron tip, melt the

Soldering Temperature for Different Components

S.No.	Electronic Component	Soldering Temperature	Soldering Time
1.			
			-
	Electrolytic capacitors		
	Silicon diodes		
6.			
7.			

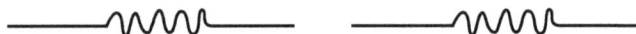

—————∿∿∿————   ————∿∿∿————

8.			
	Power Transistors		
11.			
			-

Chapter-5

Concept Building Blocks

Direction of Electric Current : Conventional Current

Let us take a copper strip as shown in the figure 5.1. The copper strip has a large number of free electrons. When electric pressure or voltage is applied, the free electrons, being regatively charged, will start moving towards the positive terminal around the circuit as shown in the figure (5.1). This flow of free electrons is called electric current.

The actual direction of current (i.e. flow of electrons) is from negative terminal to the possitive terminal through the part of the circuit external to the cell. However, according to the electron theory, it was assumed that current flows from positive terminal to the negative terminal of the cell via the circuit. It was believed that electronic current was the movement of positive electricity from the positive terminal to the negative terminal. Now, it is well established that electric current is the movement of free electrons which bear negative charge.

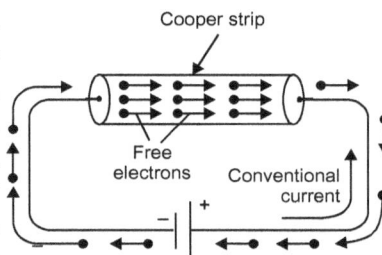

Fig : 5.1

Electric Potential : Definitionof 1 Volt.

A charged body has the capacity to do work by moving other charges either by attraction or replusion. So, the capacity of a charged body to do work is called electric potential.

Electric potential,

$$V = \frac{\text{Work done}}{\text{Charge}} = \frac{W}{Q}$$

Unit : Joule/Coulomb or volt

1 Volt Potential

A body is said to have an electric potential of 1 volt if 1 Joule of work is done to give it a charge of 1 coulomb.

Thus, when we say that a body has electric potential of 5 volts, it means that 5 Joules of work has been done to charge the body to 1 coulomb.

Maintaining Potential Difference

The device that maintains potential difference between two points is said to develp electromotive force (e.m.f.). An example of a cell. (Fig. 5.2) shows the familiar voltaic cell. It consists of copper plate (called anode) and a zinc rod (called cathode) immersed in dilute H_2SO_4.

rod. This transference of electrons takes place through the agency of dil. H

Q coulombs

Q coulombs. The chemical action of the cell did a certain amount of work (say W

the two plates will be W/Q

will be attracted through the wire to copper plate. The chemical action of the cell now transfers an equal amount

through the cell to maintain original potential difference (*i.e.*, W/Q). This process continues so long as the circuit is

copper plate is the electric current.

same thing.

Concept of E.M.F. and Potential Differences

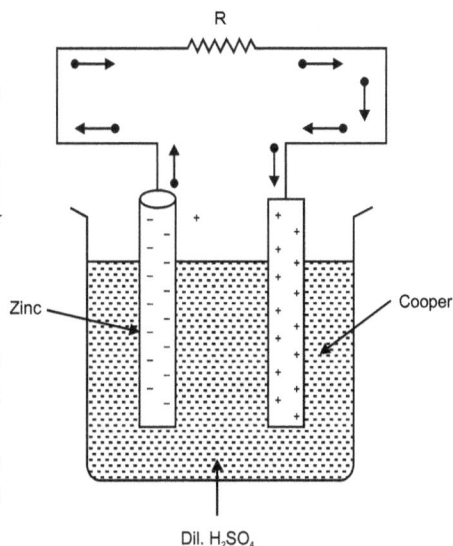

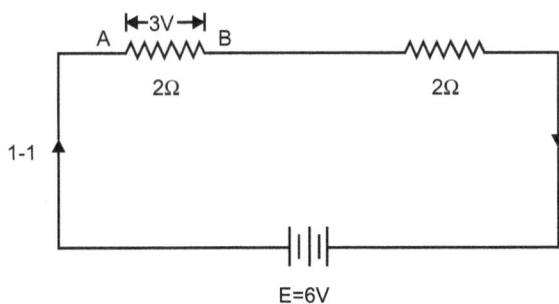

it has lost all its energy originally supplied by the battery. The battery now supplies fresh

Internal Resistance of Supply

across the terminals of cell, the current I

(a)   (b)

resistance r of the cell so that terminal

relationship between E and V can easily be

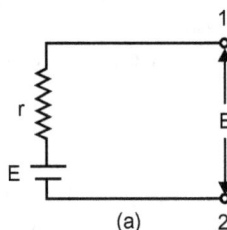

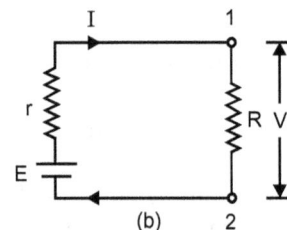

$$I = \frac{E}{R+r}$$

or $IR = E - Ir$

$IR = V$

$V = E - Ir$

Electrical Energy

The total work done in an electric circuit is called electrical energy.

Electrical energy, $W = I^2 Rt = \frac{V^2}{R}t$

power is supplied for 1 hour.

The electricity bills are made on the basis of total electrical energy consumed by the consumer.

Energy Formulas

It has already been discussed that electric power as well as electrical energy consumed can be

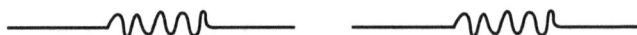

(i) Electric power, $\qquad P \quad I^2 R \quad \dfrac{V^2}{R}$ watts

Electrical energy consumed, $\qquad W \quad I^2 Rt \quad \dfrac{V^2}{R} t$

(ii) Electric power, $\qquad P \qquad VI$ watts

Electrical energy consumed, $\qquad W \qquad VIt$

These formulas apply to any type of load including the one mentioned in point (i).

Open Circuits

result of component failure or disintergration of a conducting path such as the breaking of a wire.

Short Circuits

to insulation failure, shorts of components etc.

Kirchhoff's Laws

that electric theory is based.

First law

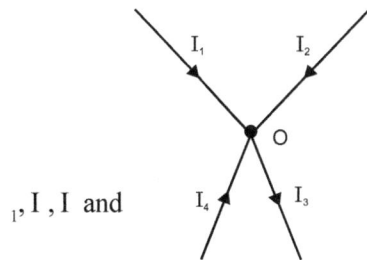

$_1$, I , I and

I

$(I_1 \quad I_2 \quad I_4 \quad -I_3$

or $\quad I_1 \quad I_2 \quad I_4 \qquad\qquad\qquad I_3$

$\qquad\qquad\qquad\qquad\qquad\qquad\qquad\qquad\qquad$ $_1$, I and I

$\qquad\qquad\qquad\qquad$ 1 $\qquad\qquad\qquad\qquad\qquad\qquad\qquad$ 1 $\qquad\qquad\qquad$.

Second Law :

In any closed circuit or mesh,

circuit and go back to that point after going round the circuit, there is no increase or decrease in

Signs of E.M.Fs. and Voltage Drops

branch.

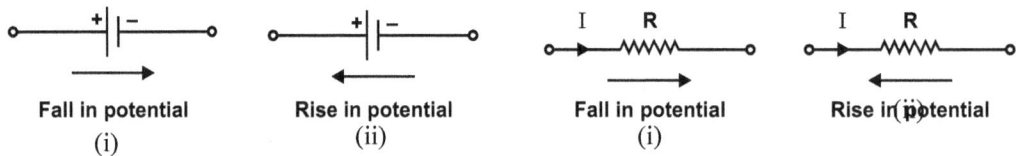

Fall in potential (i)   Rise in potential (ii)   Fall in potential (i)   Rise in potential (ii)

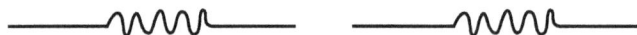

independent of the polarity of the e.m.f. in the circuit under consideration.

Ideal Voltage and Current Sources

sources.

Ideal voltage source

that maintans a constant terminal

is drawn from it.

(i)   (ii)

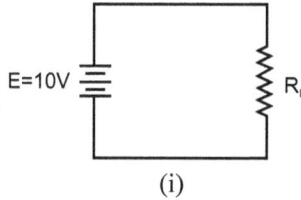

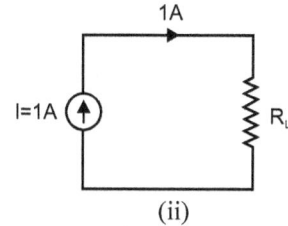

Ideal current source :

current to any resistance connected across its terminals.

, the

Real Voltage and Current Sources
Real voltage source

when current is drawn from it.

with a resistance equal to its internal resistance (R_{int}

This representation can be used to

from it. Note that internal resistance is an inherent property of a source, it is not a discrete component that can

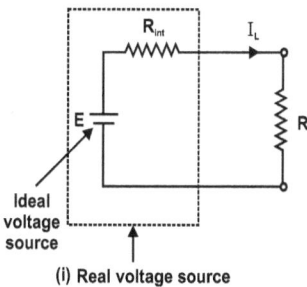

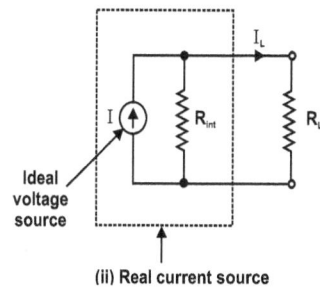

(i) Real voltage source   (ii) Real current source

int

Real current source

resistance (R_{int} is connected across the terminals, the current (I) produced by the source and R_{int}

R_{int} becomes greater, the current source approaches the ideal current source more closely.

Source Conversions

Real voltage source
(i)

Equivalent current source
(ii)

R_{int} is the same but in parallel with the current R_{int}

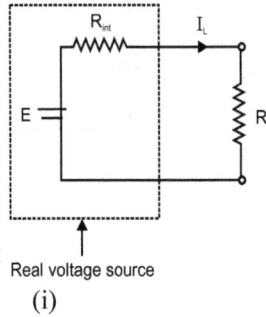

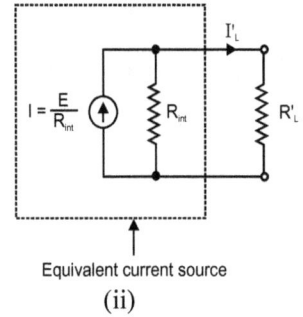

I_L	$\dfrac{E}{R_{int}+R_L}$
I'_L	$I \times \dfrac{R_{int}}{R_{int}+R_L}$
	$\dfrac{E}{R_{int}+R_L}$
I'_L	I_L

How Does a Capacitor Store Charge?

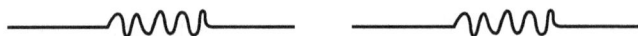

This shows that a capacitor stores charge.

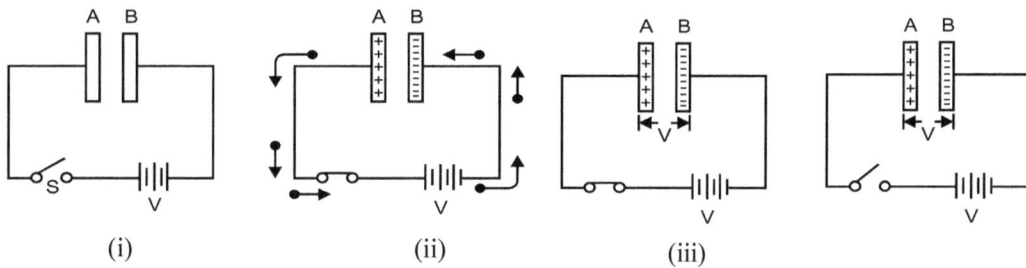

(i)   (ii)   (iii)

Uniform and Non-uniform Magnetic Field

Uniform magnetic field

The mag

points in the region.

paper is represented by parallel and equidistant magnetic   (i)   (ii)   (iii)

field

acting perpendicular to the plane of the paper and directed downwards

the paper and directed upwards is represented by equally spaced dots as

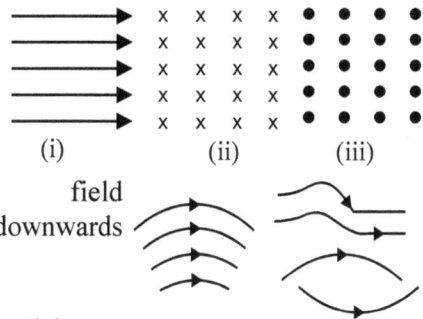

Non-uniform magnetic field :

For example, the magnetic field due to a bar magnet is non-uniform i.e. magnitude and direction of magnetic flux density varies from point to point. A non-uniform magnetic field is represented by converging or diverging magnetic lines of force. Fig. 5.12 (iv) shows magnetic lines of force due to non-uniform magnetic field where direction of magnetic field varies from point to point. In fig. 5.12 (v), both magnitude and direction of magnetic field are not constant.

❐

Project-1

Short Circuit

Introduction

how hot the wire gets.

Materials Required

Assembly

from the terminal as possible.

To Do and Notice

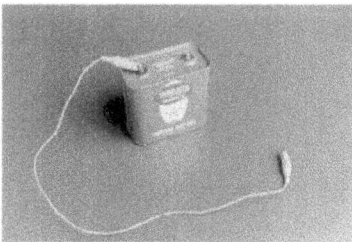

One strand of umbralled
picture-hanging wire

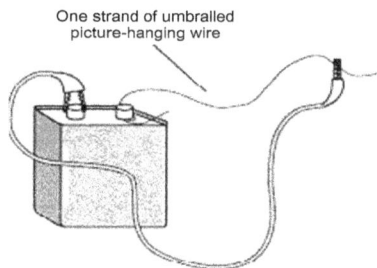

What Happens?

The thin iron wire is a good conductor of electricity, but not as good as copper wire, which is

In a normal electric circuit, an electric current powers appliances such as tubelight, refrigerator or

bypass the appliance on a path that has little or no resistance, for example, when frayed insultion

fuse contains a thin strip of wire, somewhat like the thin iron wire in our experiment. The current that goes to appliances must also pass through this strip wire. If a short circuit occurs or if too

Try It Yourself

the striking colours made by the alumunium oxide layers formed when the aluminum gets hot.

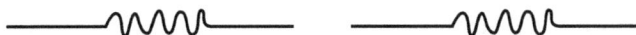

———⌇⌇⌇———   ———⌇⌇⌇———

Project-2

Bread Board or Plug-in-Board

How will you connect an LED in a breadboard?

Introduction

A breadboard or protoboard is usually a construction base for prototyping of electronics. The terms 'bread board' is commonly used to refer to a solderless bread board (plug board). In this experiment you will test how the plug-in board works by lighting up the red lamp provided. It is referred to as a 'Light emitting diode' or LED.

Materials Required

Plug-in-Board, one 330 Ω resistor, 9V lattern battery, connecting wire, an LED and a wire.

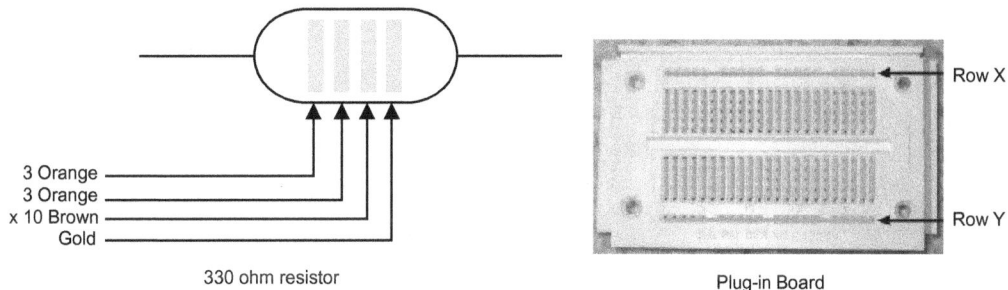

3 Orange
3 Orange
x 10 Brown
Gold

330 ohm resistor

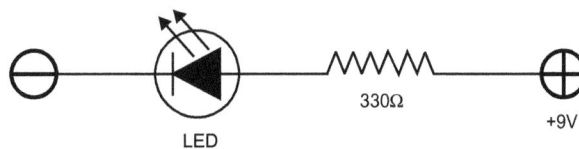

Row X

Row Y

Plug-in Board

Assembly

(1) Plug in 9 V battery, red wire into row X and black wire into row Y.

(2) Plug in 330 Ω resistor to pinholder 1-X and 1-C.

(3) Plug in the LED to holes 1-E and 1-F (longer leg of the LED should be in 1F)

(4) Plug in jumper wire from 1-G to 1-Y.

Fig. 2.1 : Bread Board

To Do and Notice

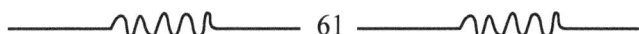

LED

330Ω

+9V

Fig. 2.2 : Circuit diagram

What Happens?

causing it to burn out.

Try It Yourself

Hint:

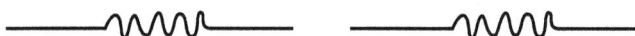

Project-3

Transitor Amplifier

How does a transistor amplify a voice signal?

Introduction

The process of raising the strength of a weak signal without any charge in its shape is called amplification. (A device which amplifies the input signal is called the amplifier. An electrical signal can be amplified by using a device which allows a small current or voltage to control the flow of more current from a DC power source.) Transistors are the basic device which provides control of this kind. There are two general types of transistors, bipolar and field-effect. The difference between these two types is that for bipolar devices an input current controls the more flow of current through the device, while for FET an input voltage provides the control.

Materials Required

Batteries, NPN transistor, microphone, loudspeaker, resistors, connecting wires etc.

Assembly

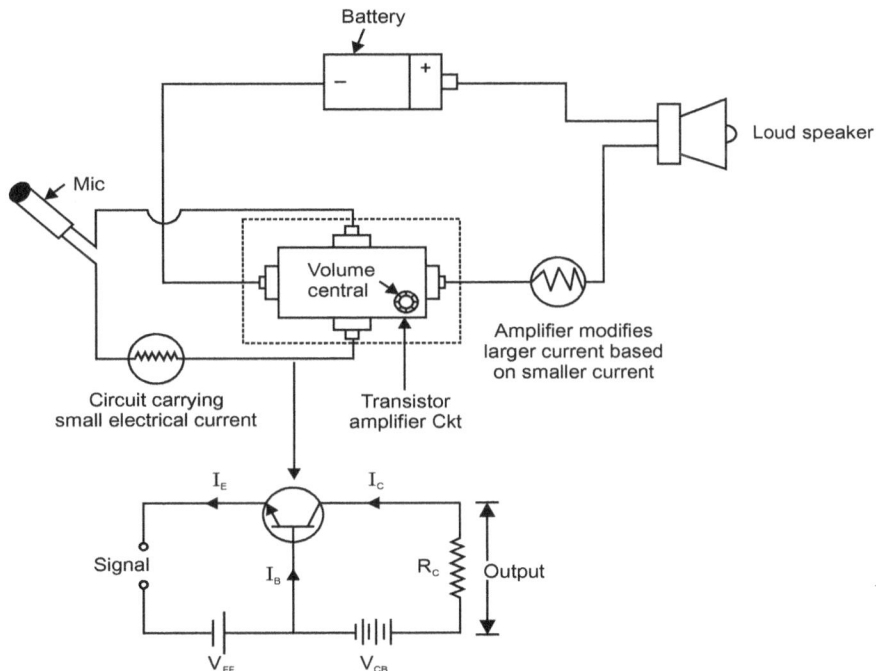

Fig. 3.1 : Circuit Diagram of Transistor Amplifier

$_{EE}$ is applied in the input circuit in addition to the signal as shown.

To Do and Notice

electrical signal is our original (input) signal. The input signal goes into the transistor and it

into the loudspeakers.

What Happens?

an appreciable change in emitter current. This almost causes the same change in collector current

Try It Yourself
Take a PNP transistor instead of NPN transistor and repeat the same experiment.

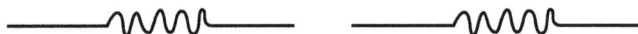

Project-4

Electromagnetic Effect

Can a current-carrying wire behaves like a magnet?

Introduction

A permanent magnet is an object made from a material that is magnetized and creates its own persistent magnetic field. When a conductor carries a current, it creates a magnetic field around it. The direction of such magnetic field depends on the direction of the current passing through the conductor. So, electric current and magnetism are very closely related to each other. This relationship plays an important role in the DC machines.

Materials Required

A plastic/wooden ruler, compass needle, one long nail, 100 cm insulated connecting wire, 6V battery.

Assembly

Take an iron nail and wind 100 cm insulated copper wire over it; then connect its free end to a 6 V battery and place it on the ruler facing N-S direction. Place a compass needle near the iron nail wounded with copper wire. Observe the compass needle.

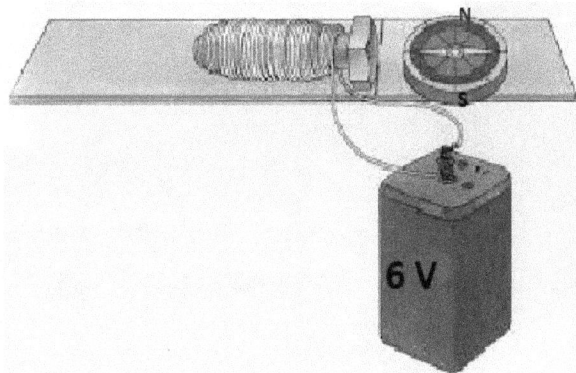

To Do and Notice

It is observed that the compass needle will get deflected (towards west) on changing the direction of current, the deflection of the compass needle gets reversed also.

What Happens?

on any magnet.

Try It Yourself

Instead of iron bar, take a non-conducting material bar as dry wood or plastic and repeat the same

Project-5

Electromagnetic Force

Can an electromagnetic force power a motor?

Introduction
When a current flows through a conductor, it behaves like electromagnet and creats a field around it. This field is time-varying containing both electric and magnetic fields that feed upon each other even in vacuum. Due to this electromagnetic field, we experience the electromagnetic force. Electrical motors work due to the combination of electricity and magnetism.

Materials Required
Two bar magnets with marked poles, A cotton reel, A few meters of copper wire. Three pieces of electric cable, a wooden skewer type toothpic, two metal drawing pins, a small piece of wood, a paper clip, two elastic bands, four large corks, two iron washers, a 9V battery and sticky tape.

Assembly
1. Wind the copper wire a number of times around the cotton reel from the bottom to the top. Wind as tight and closely as you can, leaving the two ends of the wire free. Keep the wire around the reel with elastic bands.

2. Thread the toothpick through the holes in the cotton reel, taking care not to pierce the copper wire if you possibly can. Thread a washer at either end of the stick. Thread an end of wire through each washer.

3. With sticky tape, fix the magnets to two corks and place them with the opposite poles facing each other. Place the other two corks crossways between the two magnets. Put the stick on top of these corks and fasten down with sticky tape.

4. Stick the two drawing pins in the wood about 2 cm apart. Open up the paper-clip and thread one end under the drawing pin so that the clip can be turned to connect up with both pins. This will be your switch wire.

5. Peel off a little of the plastic covering from the ends of the three pieces of electric cable. Then make this circuit. Keep one end of wire between a battery contact and the washer, the other end from the other washer to the drawing pin on

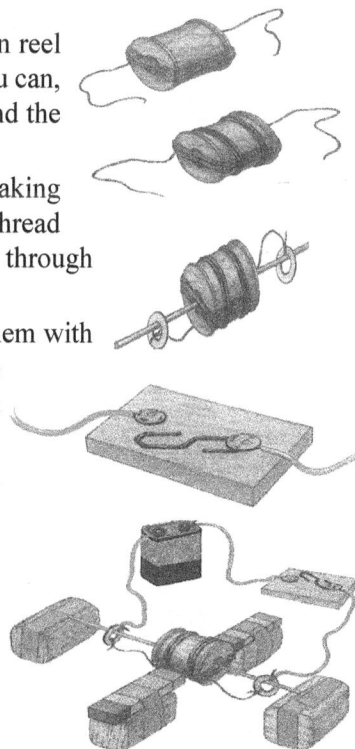

the wood; and one from the drawing pin to the second battery contact.

6. Turn on the switch by putting the paper clip on the drawing pin so that the current of

To Do and Notice

What Happens?

Try It Yourself

Project-6

Wiring of a House

Can you assemble a household circuit comparing three bulbs, three (ON/OFF) switches, a fuse and power source?

Introduction

The permanent cabling and fittings in your home are called electrical installations or wiring of your home. In this process fuses and circuit breakers are used. The switchboard protects the electrical wiring system in your home. It comprises the main power switch alongwith a number of fuses or circuit breakers. The main difference between a circuit breaker and a fuse is that a circuit breaker can be reset whereas a fuse once blows must be replaced. All home appliances are connected in parallel with each other so that they get same operating voltage.

Materials Required

Three bulbs (20 W, 50 W and 100 W), three (ON/OFF) switches, flexible connecting wire with red and black plasting covering, a fuse wire, a two pin plug. Main electric board with a two-pin socket and main switch.

Assembly

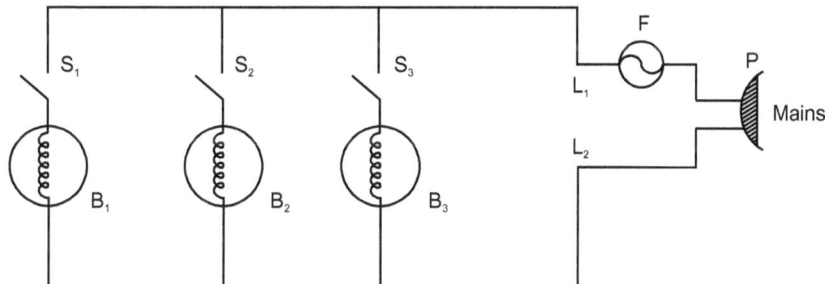

1. Connect one end of the bulb-holder to the red flexible wire through switches in series. Connect the other end of the bulb-holder to the black flexible wire.
2. Connect the three bulb-switch combination in parallel, red wire ends at one point and the black wire ends at the other point.
3. Take two long flexible wires to serve as lead wire; one wire is red and the other is black.
4. Connect the red wire ends to the red lead wire L_1. It will serve as a lower lead.
5. Connect the black wire ends to the black lead wire L_2. It will serve as neutral-lead.

in lower hole of the socket.

To Do and Notice

Make the switches   one by one. Then put them   one by one.
The bulb glows when the switch is made on. It stops glowing when the switch is put   .

What Happens?

to do action.

Try It Yourself

well as in parallel and see which bulb glows more brighter in which combination.

Project-7

Measurement with Multimeter

What do you measure with a multimeter?

Introduction

A meter is a device that measures electrical quantities in your electronic projects. A multimeter therefore, is a combination of different types of meters all in one box. It is a single measuring device that acts as an Ammeter, a Voltmeter and an Ohmmeter. For this reason, it is also called AVO meter.

We use a multimeter for the following purposes :

(1) To measure resistance, voltage (AC/DC), current (AC) and to check continuity of a given circuit.

(2) To see unidirectional flow of current feature in case of a diode and an LED (Light Emitting Diode)

(3) To check whether a given electronic component (e.g. diode, transistor or IC) is in working order.

Materials Required

A multimeter (here analog multimeter), connecting wires

Assembly

Measurement of resistance

Turn the multimeter knob to ohmmeter range of middle value. Touch the two lead points with each other and adjust the pointer to zero reading on resistance scale (maximum deflection). The zero lies on the extreme right of the multimeter scale.

Now touch both the leads with the two ends of the resistor whose resistance is to be measured. The position of the pointer on Ohmmeter scale gives the value of the resistance (decrease value of the range to get bigger deflection and accurate result.

Measurement of voltage (AC/DC)

Turn the multimeter knob to voltage range. Touch the two leads with the two points between which the voltage (potential difference) is to be measured. The position of the pointer on the scale gives value of the voltage (change the range to get maximum deflection).

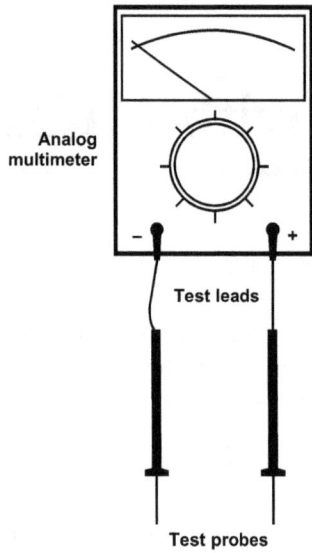

Analog multimeter

Test leads

Test probes

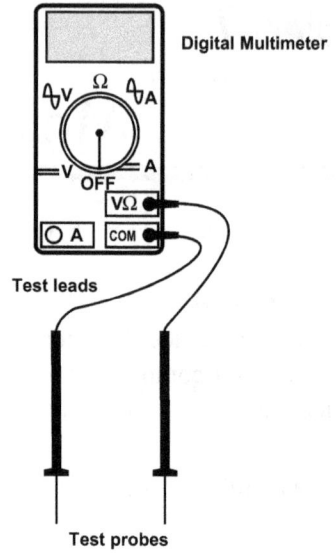

Digital Multimeter

Test leads

Test probes

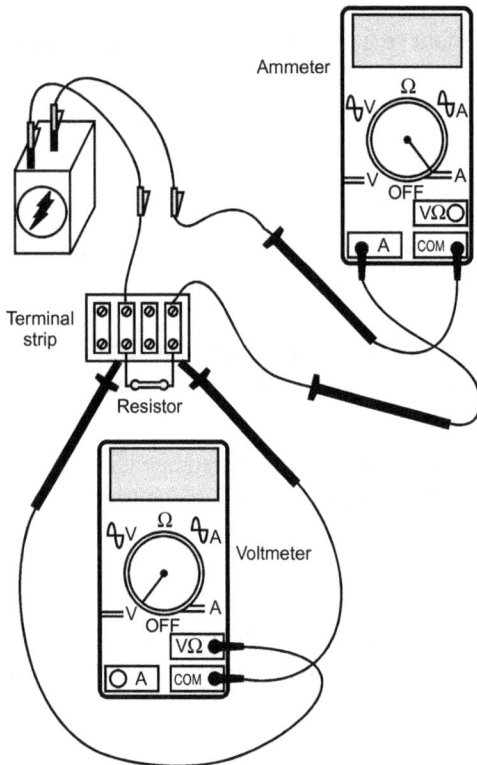

Ammeter

Terminal strip

Resistor

Voltmeter

6-volt "lantern" battery

1.5-volt "D-cell" battery

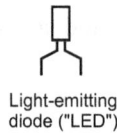

Light-emitting diode ("LED")

Permanent magnet motor

Measurement of current (AC)

Checking the continuity of a circuit

lies on the extreme right of the multimeter scale.

Now, touch the two leads with the two points, one with each point, between which continuity

continuity.

from the multimeter cell).

Observation of unidirectional flow of current in case of diode or LED

Checking of working order of electronic components

Put these components in their normal working circuit and operate the circuit. If results are satisfactory, the components are in working order.

Try It Yourself

one.

Project-8

Resistance of a Resistor

Can you find the value of an unknown resistor by Ohm's law?

Introduction

Resistors are devices used in various technological devices in order to control the flow of current and thus make the product efficient. They can be found in any electronic equipment that uses a circuit in order to control its several functions. The strength or resistance of a resistor is measured in ohms (Ω). The value of an unknown resistor in a circuit can be calculated by ohm's law. Ohms law states that at constant temperature the current (I) flowing through a conductor is directly proportional to the potential difference (V) across the ends.

Materials Required

1. Regulated power supply (R.P.S.) (0-10V)
2. Ammeter (0-10 mA)
3. Voltmeter 0-10V)
4. Unknown resistor
5. Connecting wires

Assembly

Resistance is the property of a substance due to which it opposes the flow of current through it. Resistance obeys ohm's law.

According to ohm's law

Resistance	R	=	V/I
where	V	=	voltage across resistance
	I	=	current flowing through the resistance

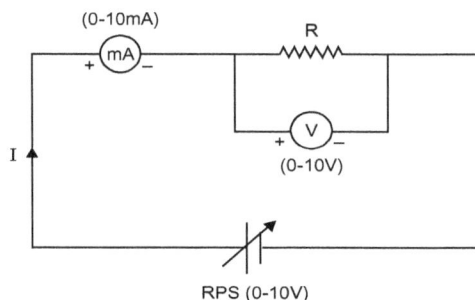

Fig. 8.1 : Circuit Diagram

the table shown below

To Do and Notice

S.No.	V (Volts)	I (mA)	R = V/I (KW)
1	1	1	1
			1
			1
			1
			1
6	6	6	1
7	7	7	1
8	8	8	1

Average value :

What Happens?

$\dfrac{V}{I}$ is always constant and this is known as resistance. The

		Red	
1			

Try It Yourself

Project-9

Inductance of an Inductor (Coil)

Can you find the inductance (L) of a given coil by V-I method?

Introduction

Inductors are sometimes referred to as 'coil' because most inductors are physically constructed of coiled sections of wire. Inductors are passive devices used in electronic circuits to store energy in the form of magnetic field. They are the compliment of capacitors which store energy in form of electric field. An inductor is characterized by its strength known as inductance. The property of a coil to due to which it opposes any increase or decrease in current or flux through it, is known as self inductance.

Materials Required

1. Auto transformer : 230V/(0-270V) -1
2. Ammeter (0-10 A) M.I -1
3. Voltmeter (0-150 V) M.I -1
4. Watt meter 150 V, 10 A, LPF -1
5. Connecting wires
6. Unknown Inductance coil -1

(Decade Inductance box)

Assembly

Inductance is measured in henry (H). The strength of a coil is quantitatively measued in terms of coefficient of self inductance L.

$$X_L = \sqrt{Z^2 + R^2}$$

Inductance $\quad L = X_L/2\pi f$

Where X_L = Inductive reactance

$\quad$ f $\quad$ = frequency = 50 H_Z

$\quad$ Z $\quad$ = Impedance of the coil = V/I

$\quad$ R $\quad$ = Resistance of the coil = W/I^2

1. Connections are made as per the circuit diagram
2. Switch ON the power supply
3. The voltage across the coil is varied in steps with the help of Auto transformer. The readings of voltmeter (V), Ammeter (A) and Watt meter (W) are noted in the table.

Watt meter
150V, 10A LPF

230V
50Hg
AC supply

10A

10A

(0-10A) MI

(0-150V) MI

230/270V
Auto Transformer

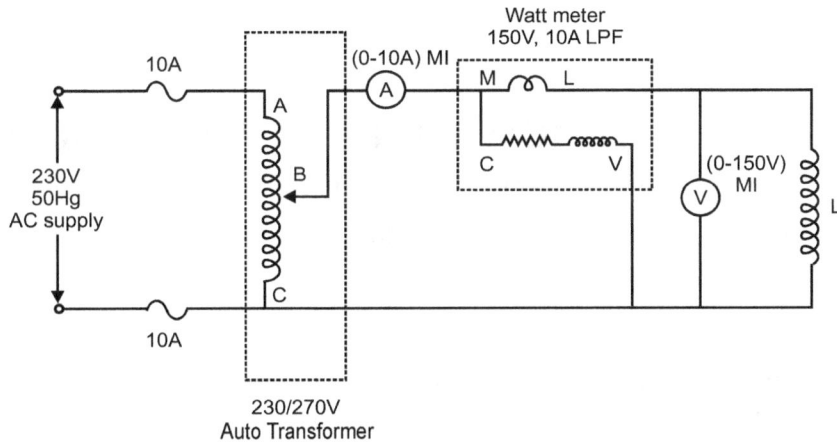

To Do and Notice

Table

S. No.	V (Volts)	I (Amps)	W (Watts)	Z = V/I (ohms)	R = W/I² (ohms)	²-R² (ohms)	L=X$_L$ (mH)
1			11				
					1.78	8.77	
		6.16		8.11			

What Happens?

Try It Yourself

Project-10

Capacitance of a Capacitor

How do you find the value of unknown capacitance in a circuit?

Introduction

The charge-holding property of a body is called capacitance or capacity of a capacitor. A capacitor essentially consists of two conducting surfaces separated by an insulating material. It has the property to store electrical energy in the form of electrostatic charge. Using De Sauty Bridge we measure an unkown capacitance in terms of a standard capacitance i.e. comparing two capacitances. Two ratio arms of this bridge consists of pure resistors and rest two consist of capacitors where one is of unknown value and the other is standard capacitor.

Materials Required

1. Audio Frequency generator
2. Decade Resistance boxes
3. Decade capacitance box
4. Head Phone set
5. Unknown capacitor

Assembly

De Sauty's Bridge can be used to measure the value of capacitor. This bridge is very simple. Let

$C_1 = C_x$ = Unknown capacitor

C_2 = A standard capacitor

R_1, R_2, R_3, R_4, are standard resistances

At bridge balance conditon.

$C_1/C_2 = R_2/R_4 = R_1/R_3$ (or) $C_x/C_2 = R_1/R_3$ and $R_2/R_4 = R_1/R_3$

Bridge balance can be obtained in two ways: (i) keeping C_1, C_2 constant and varying R_1 or R_3 (ii) keep the values of R_1, R_2, R_3, R_4, so that $R_1/R_3 = R_1/R_4$ is satisfied and vary the capacitor C_2. The value of unknown capacitance is found by using the formula.

$C_x = C_2 R_1/R_3$

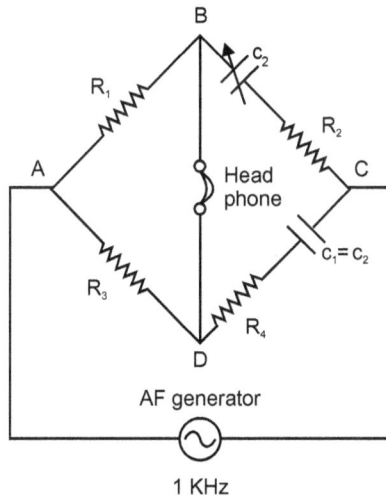

$_1$, R , R , and R so that the condition R$_1$ is

minimum sound is heared from head phones or ear phones.
$_1$, R , R , R   in the table.
$_1$, R , R , R

To Do and Notice

S. No.	R$_1$ ( )	R$_2$ ( )	R$_3$ (W)	R$_4$ ( )	C$_2$	Cx = C$_2$R$_1$/R$_3$
1						
6						

x

What Happens?

dielectric increases the concentration of electric lines of force between the plates and the charge

Try It Yourself

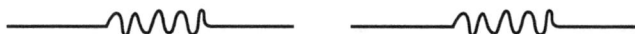

Project-11

Battery Combination

Why is the position of the cell or battery important?

Introduction
The device which supplies energy to the circuit is called battery. The flow of current generated by batteries can happen not only between poles of the opposite sign but also on the correct combination of batteries with one another. Cells and batteries may be connected in series, parallel or combinations of both. In series combination, positive terminal of one cell is connected to the negative terminal of another cell. Cells or batteries connected in parallel have their like terminals connected together.

Materials Required
Two 1.5 volt batteries, a small bulb, two pieces of electric cable with the ends uncovered, A ruler and sticky tape.

Assembly
1. Stick the two batteries along the length of the ruler with the sticky tape. Follow a positive pole (marked at the positive end with a plus (+) sign and a negative pole marked at the opposite end with a minus (−) sign.
2. Use the sticky tape of stick at the ends of two wires to the opposite heads of the batteries. Bring the two batteries together. Then touch the bulb with free ends of the wires, as shown.
3. Now, reverse the position of the batteries, so that the two positive poles touch.
4. Re-connect the wires to the two ends of the double battery and touch the bulb again,

To Do and Notice
The bulb does not light up.

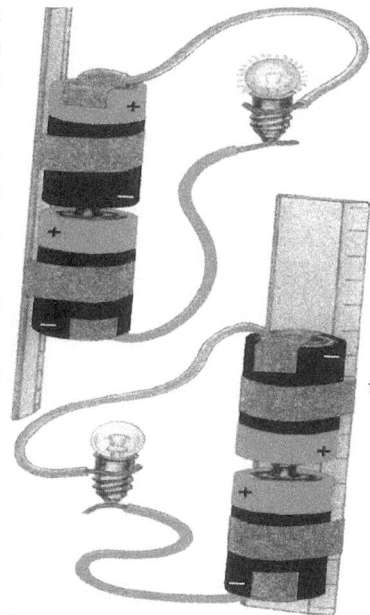

Fig. 11.1 : Circuit Diagram

What Happens?

In case I when we touch the bulb with free ends, the bulb lights up because the electrons generate

battery. Something happnes if two batteries are connected to each other because the electrons still

work if a battery is inserted with the poles in wrong positions.

Try It Yourself

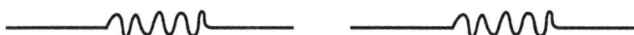

Project-12

Electrical Machines

How do you bind a powerful DC Motor?

Introduction

In all rotating electrical machines an electromechanical energy conversion takes places. In all electrical machines, a charge in flux is closely associated with the mechanical motion to cause electromechanical energy conversion when electrical input energy is converted to a mechanical energy, the machine is called a motor. In simple words a device that converts direct current into mechanical energy is called DC motor. It is based on the principle that when a current-carrying conductor is placed in a magnetic field, it experiences a mechanical force whose direction is given by Fleming's left-hand rule and whose magnitude is given by gauge newton.

Materials Required

- Steel road 1/8" × 7" long
- 8 d common nails – 2½" long flat head type
- Nearly 20 ft., 22 gauge magnetic wire
- 35 mm film canister, rounded with cover
- Ply board 6" × 6" piece
- Two plywood boards 3" × 1" × (½)" pieces each
- Two fender washers, 1" diameter × 1/8" hole
- Steel pipe strap, 12(½)" long
- Two strong ceramic magnets
- Standard hook-up wire, 16 inch, 12 gauge
- Two alligator clip test leads
- 6 V lantern battery
- Duct tape, electrical tape, sandpaper
- Mini hot melt glue sticks
- Hand raw
- Drill
- Wire Stripper/Cutter with 14 gauge hole (to cut and strip wires)
- Aviation strips (to cut pipe strap)
- Bolt cutters (to cut steel rod)
- Pliers (to bend pipe strap)

Mini hot melt glue gun
Ruler
Marker

Assembly

1. **Making the armature - An electromagnet on a stick**

that their opposite ends touch each other.

(ii) Stick the steel rod between the nails sliding it exactly at the centre of the nails and glue in place with hot glue.

the right of the steel rod, and wrap the

the wire back and forth staying on the right side of the nails until you get to the

around and continue wrapping of the

free at the end of the coil.

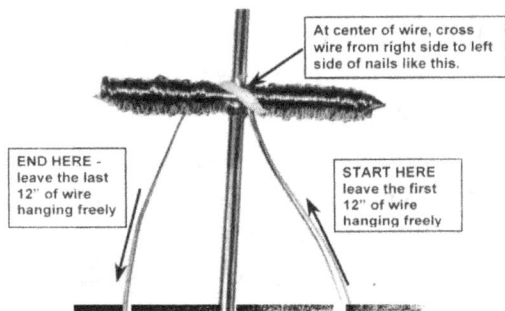

The two ends of the wire coil that you left while wrapping are called electromagnet lead, because they lead electricity into and out of the coil.

bare copper.

of the electromagnet up to the same pole of a permanent magnet. If one end of the electromagnet feel attracting and other feel repelling. The repelling force is always weak; if it is not so, it means the turns of wire are not going in the same direction. Now carefully unwarp the wire of one side and rewarp it again with great care.

Making the commutator :

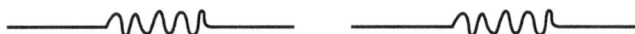

Electrical Machines

(i) The commutator in this motor is made is by

touch the brushes, which are connected to the

coil, it becomes an electromagnet.

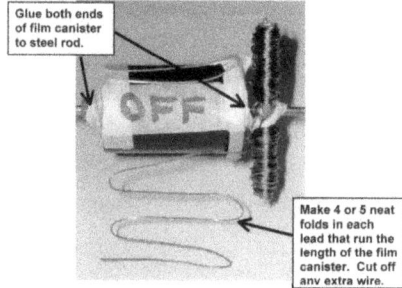

(iii) Slide the canister onto the steel shaft until it is

and hot glue it at both ends.

the sides facing you and on the opposite side. These

the leads to get electricity into the electromagnet, so

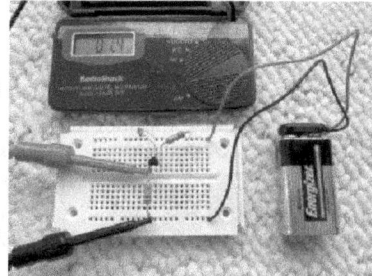

Making the bearing blocks :
bearings on both ends to hold the shaft in place and

the corners.

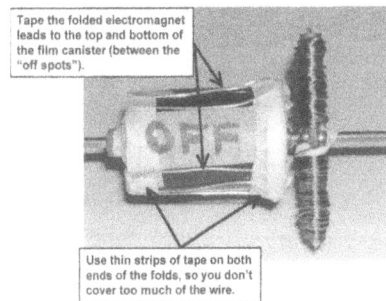

Making the Brushes : The brushes in a motor are attached to battery and they are used

(i) Take two pieces of the standard wire, each 8 inches long

are tight. These ends of the brushes will be connected to the battery using the alligator clip test leads.

armature spins freely and easily in the brearing blocks before gluing the other one down.

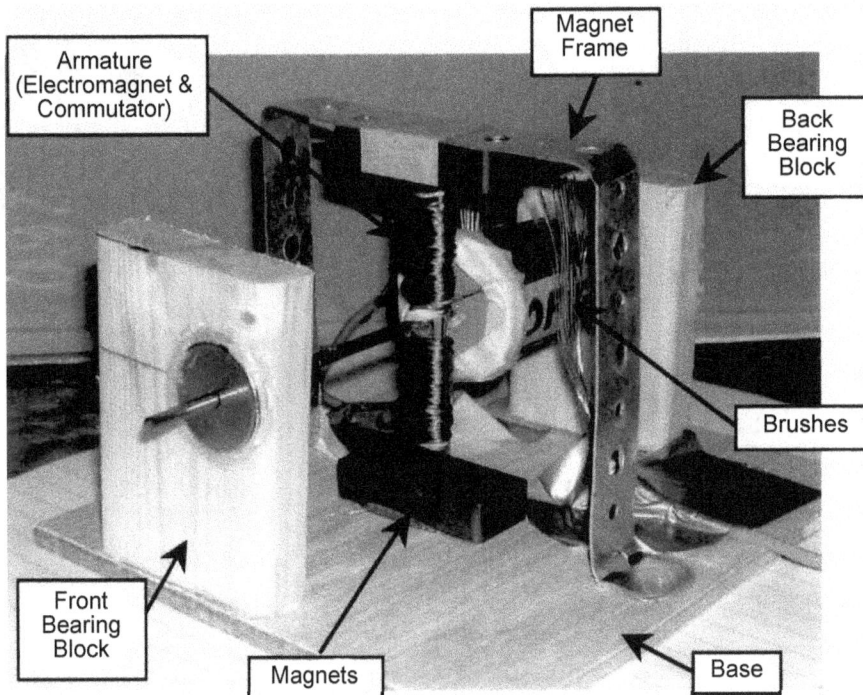

To Do and Notice

What Happens?

direction while all the conductors under S-pole carry currents in the opposite direction.

Try It Yourself

Hint :

ceiling fan.

Project-13

Water Sensor

How do you make water sensor alarm system?

Introduction

A sensor is a converter that measures a physical quantity and converts it into a signal which can be read by an observer or by an electronic instrument. A sensor's sensitivity indicates how much the sensor's output changes when the measured quantity changes. In this project, we use a transistor to amplify the signal from humidity (water) sensor.

Materials Required

30 pF capacitor (water-vapour sensor) two N-P-N transistors, two 100 KΩ, one 300Ω and one 330Ω reistors, 9 V lantern battery, plug-in board, connecting wires, battery clip, red LED, 100 KΩ, variable resistor, screw driver and auto range digital multimeter.

Assembly

Humidity sensor works through electrical conductivity, when water vapour condenses on the plates, it forms a thin film that conducts electricity between the two plates. In that case the water sensor is like a resistor except its resistance value changes depending on the amount of water in the air. It is made by cliping the top of a 30 pF capacitor using a pair of wire clippers.

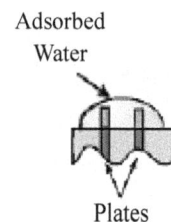

Fig. 13.1   Fig. 13.2

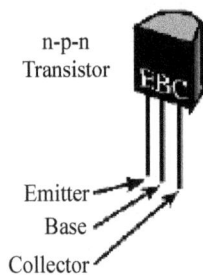

n-p-n
Transistor

Emitter
Base
Collector

Fig. 13.3

C
B
E

Symbol

Fig. 13.4

Transistor connections:
Our kit contains a N-N-P transistor which has emitter, base and collector terminals. The collector and emitter are connected to a pipe that carries water and base connected to faucet (a device that controls flow of water from the pipe).

Ordinarily the base is closed and the resistance between collector and emitter is very large. When a small voltage is applied to the base, it reduces the resistance between the collector and emitter, allowing the flow of current.

Wiring up the water sensor: The circuit diagram is shown at right and experimental arrangement of the water sensor in the above figure.

Test the circuit with the multimeter

setting, touch the sensor with your

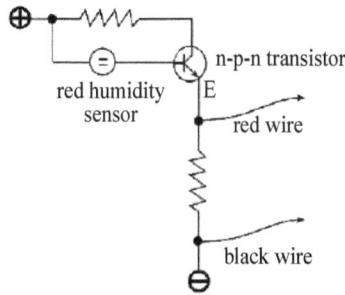

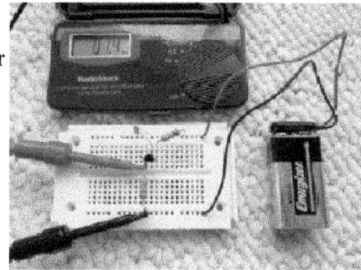

reading indicates a change in the humidity of the air around the

Adding over limit Alarm

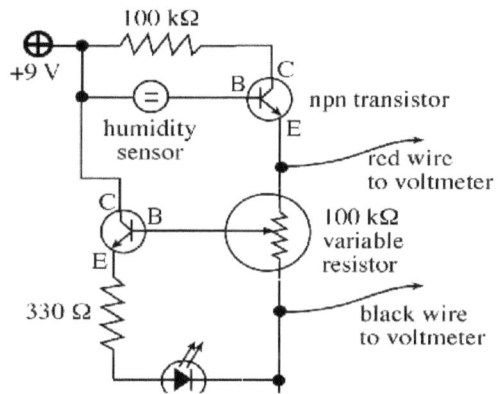

To Do and Notice

What Happens?

Try It Yourself

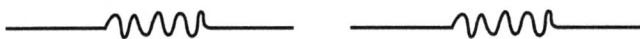

Project-14

Electronic Door Bell

How do we construct a 9-V Door Bell Chime using 555-timer?

Introduction

This door bell chime project is very low cost and affordable project that every beginners to electronic design can do. This is very simple wired door bell alarm that you can place on the door of your house. In most wired systems, a button on the outside next to the door, located around a height of the doorknob, activates a signaling device (usually a chime bell) inside the building. Pressing the door bell button, a single-pole single-throw (SPST) push button switch momentarily closes the door bell circuit and produces a signal from 555 timer at pin-3.

Materials Required

Label		Description
U_1	:	555-timer
R_1	:	15Ω ¼W, 5% resistor
R_2	:	15KΩ ¼W, 5% resistor
R_3	:	39KΩ, L4/4W resistor
R_4	:	68KΩ, L4/4W resistor
R_5	:	10KΩ, L4/4W resistor
R_6	:	10KΩ, L4/4W resistor
D_1, D_2, D_3	:	diode IN4148
E_1	:	47µF/25V electrolytic capacitor
E_2	:	470µF/25V electrolytic capacitor
E_3	:	22µF/25V electrolytic capacitor
C_1	:	15nF/25V ceramic capacitor
Q_1	:	NPN transistor 2SC945
Speaker	:	8µ speaker
S_1	:	Normally open-push button switch
BAT	:	9V batterry with battery holder

Assembly and Theory

This project is a wonderful application of 555-timer that is configured here in an astable mode. This project uses a 555-timer integrated ciruit. So, first of all understand the 555-timer.

capable to produce highly accurate and stable delays or

transmitters, burglar alarms, etc. In addition to looking at its internal architecture, we will also consider two basic operational modes i.e. as monostable and astable

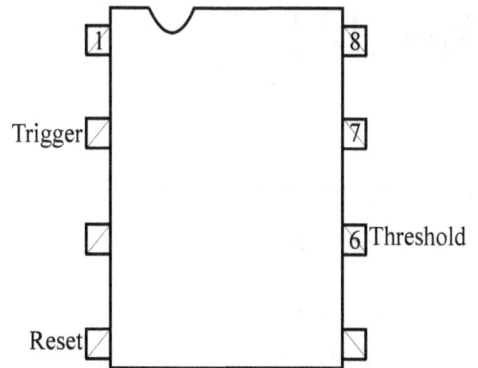

Trigger

Reset

1

8

7

6 Threshold

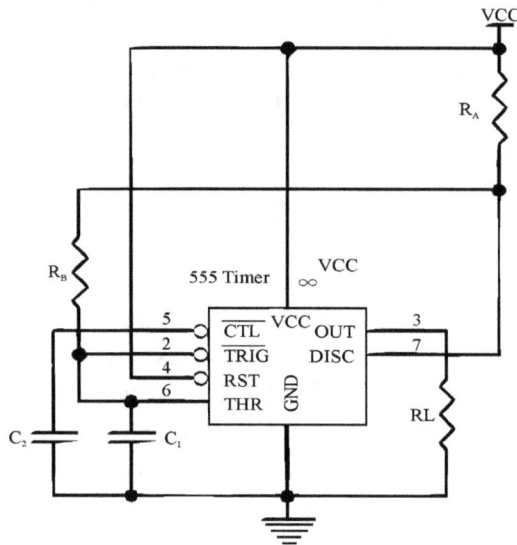

VCC

R_A

R_B

555 Timer

VCC
∞

5 CTL VCC OUT 3
2 TRIG   DISC 7
4 RST
6 THR GND

RL

C_2 C_1

and R and

. Its output triggered continuously. The result of the output is a stream of clock

to the I   charges towards V   through resistors (R   ).

─

now charges through R and R

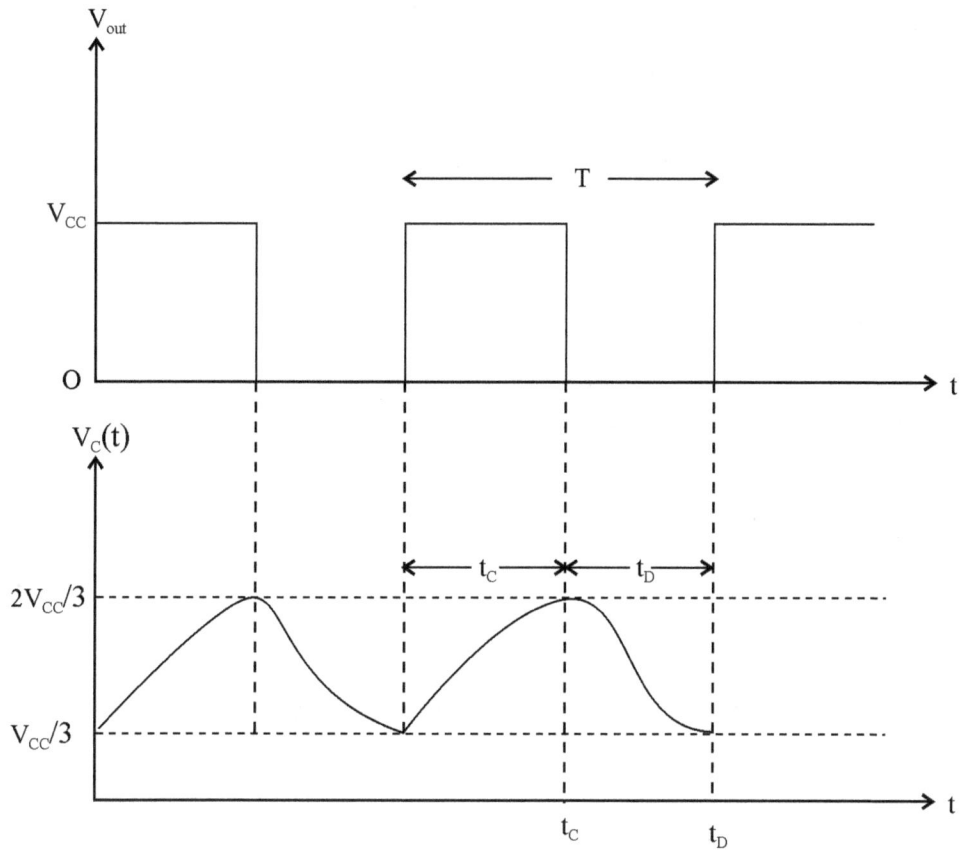

capacitor charges from V ꜀

and

t   1

t   1

1

f   1/T   ──────
               +

or   f

Here   R

R

$\quad$ 1

f $\qquad\qquad$

$\qquad$

f

$$\frac{R\ +R}{+}\ \underline{\quad}$$

T_1 $\qquad$ 1

T $\qquad$ 1

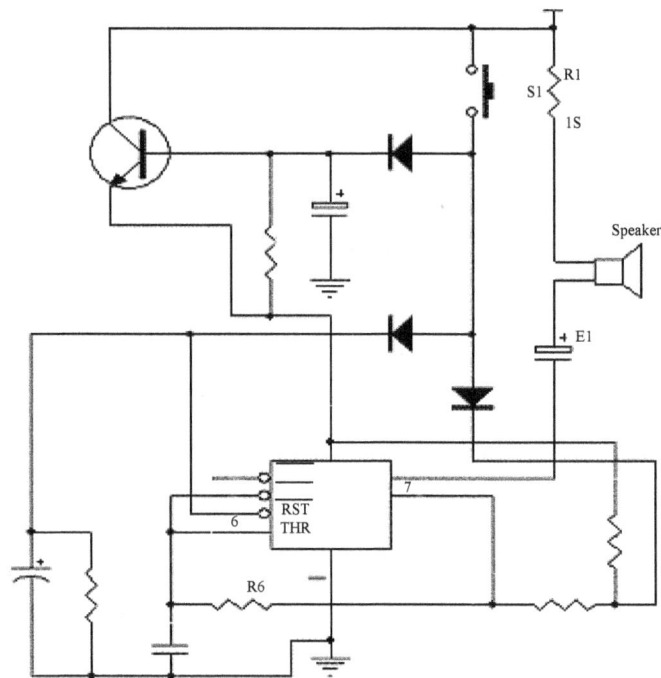

To Do and Notice

1

What Happens?

1 charges towards V   through the resistors (R

—

Try It Yourself

———〜〜〜———   ———〜〜〜———

Project-15

Electrical Conductivity

Can water conduct electricity?

Introduction

Conductivity of a substance is defined as the ability or power to conduct or transmit heat, electricity, or sound'. The conductivity of water is affected by the presence of inorganic, dissolved solids such as chloride, nitrate, sulphate and phosphate anions or sodium, magnesium, calcium, iron and aluminium cations. Organic compounds like oil, phenol and sugar do not conduct electrical current very well. Pure water is an insulator, but water-containing salt is a good conductor of electricity. Conductivity is also affected by temperature; the warmer the water, the higher the conductivity. For this reason conductivity is reported as conductivity at 25°C.

Materials Required

A glass or plastic container, two small terminals (clamps), electric cable, A 4.5 V battery. A bulb, distilled water (from a garage of hardware shop), salt, wire-cutters

Assembly

1. Fill the container with distrilled water.
2. Ask an adult to cut 3 pieces of cable strip at the ends. Connect one end of 2 wires to the battery contacts, and one free end to a terminal. Connect one end of the third wire to the second terminal.
3. Place a terminal on each side of the container so that these touch the water.
4. Connect the free ends of the wire to the bulb, one wire touching the bottom of the metal screw, the other touching the side of the screw.

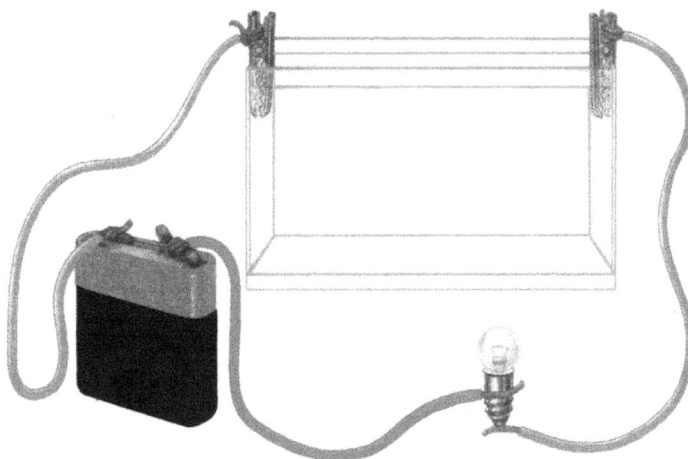

Fig. 15.1 : Circuit Diagram

To Do and Notice

Now the bulb lights up.

What Happens?

being electrically charged, start to become attracted to the terminals connected to the battery. This creates a sort of connection which closes the circuit and allows the electricity to pass through.

Try It Yourself

Project-16

Brightness of Electric Bulb

What does the brightness of light depend on?

Introduction
The brightness of a light depends on the speed with which the electrical charge flows through the conducting wires. If the path of the charges goes long, the brightness of bulb decreases. As wire goes long, the resistance of overall circuit increses so that the power loss also increases and net voltage across the terminal of bulb decreases due to decrease in this brightness.

Materials Required
A 4.5V battery, a bulb with bulb-holder, electric cable, a pencil lead and sticky tape.

Assembly
1. Connect the wire to the battery and the bulb
2. Open up the pencil. Take out the lead.
3. Fix the end of the wire to one end of the lead. Run the end of the other wire along the length of the lead.

Principle : Many electrical appliances which we use in our homes have an electrical resistance inside them, which heats up with the passage of the current and changes electrical energy into thermal (heat) energy.

To Do and Notice
When the end of the wire is run along the length of the lead, the light in the bulb varies in brightness.

Fig. 16.1 : Circuit Diagram

What Happens?
The lead offers resistance to the passage of electrical current. The longer the length of the lead which is a part of circuit, the more the energy which is absorbed, and the less the brightness of the bulb.

Try It Yourself
Take an ammeter and measure the current passing to the bulb.

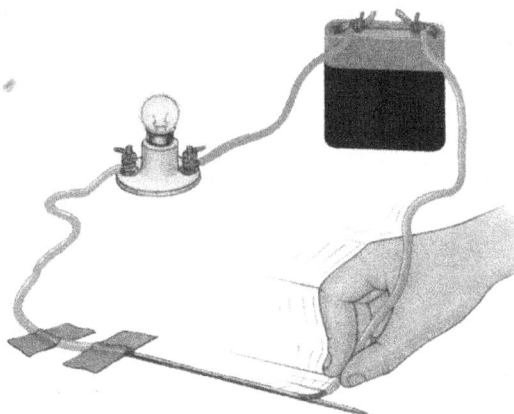

Project-17

Electrolysis of Water

What changes does electricity make in water?

Introduction

Electrolysis is the process by which ionic substances are decomposed into simpler substances when electric current is passed through them. When electricity passes through water, it can break up and change the substances which the water contains. Chemical changes occur in water and there is deposit of green substance on the bottom.

Materials Required

- Electric cable (with ends stripped)
- Two 4.5V batteries
- Two large screws
- A glass
- Water
- Salt
- An old post card

Assembly

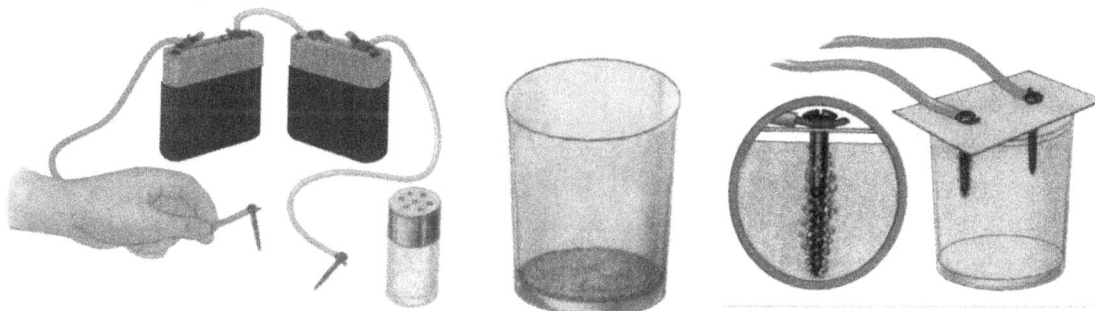

1. Connect the batteries with a piece of wire. (Negative and positive poles must not be together)
2. Connect each of the outer battery contacts to a screw as shown
3. Fill the glass with water. Add the salt
4. Make two holes in the card, a short distance apart. Place the card on the glass. Thread the screws through the holes.

To Do and Notice

What Happens?

The passage of electrical current in water causes chemical changes. The current breaks up the water (the bubbles are full of hydrogen, a component of water.) The electricity also changes the salt and the iron from the screws into other substance. This is the reason for discolouration of the water and the deposit on the bottom of the glass.

Try It Yourself

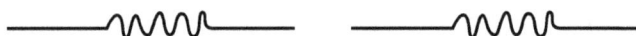

Project-18

Circles of Magnetism

How to make a magnetic field that's stronger than that of the Earth?

Introduction

Compass needles are little magnets that are free to rotate. Compass allows us to observe the direction of a magnetic field. Normally, they respond to the earth's magnetic field, orienting themselves parallel to magnetic field lines.

If we create a magnetic field that is stronger than the field of the earth, for example, by using electric currents, a compass needle will orient itself parallel to the new field.

Materials Required

- A 6 or 12 volt lantern battery
- A 1 foot (30 cm) length of heavy wire that is rigid enough to stand by itself. (You can use the wire from a coat hanger).
- A tinkertoy set for building the stand (or another improvised stand).
- A flat, rigid support surface measuring approximately 6 × 6 inches (15 × 15 cm). This can be made of posterboard or even a manila file folder). It should have a hole in the centre of it that is large enough for the wire to pass through.
- 4 or 6 small compasses, measuring about 1 inch (2.5 cm) in diameter.
- 2 electrical lead wires with alligator clips at both ends (available at radio shack).
- Adult help

Assembly

Construct a tinkertoy stand (or the equivalent), and lay the flat support surface in position on the stand. (see the diagram fig. 18.1).

If the coat hanger wire is painted or varnished, scrape the coating off to expose about 1 inch (2.5 cm) of bare metal at each end.

Insert the wire through the hole in the flat support surface, and support the wire vertically in the stand as shown in the diagram.

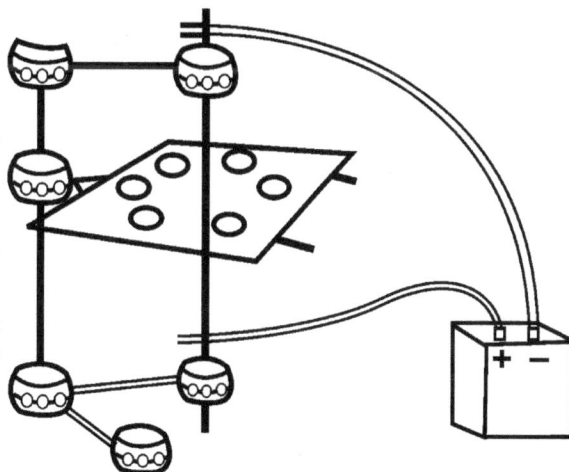

Fig : 18.1 : Circuit Diagram

lead to each battery terminal but do not attach the other ends of the lead wires to the coat hanger wire yet.

To do and notice

happens to the compass needles as the current passes through the wire.

If the electrical current is large enough, each compass will point in a direction tangent to a circle

What Happens?

of compass points towards the magnetic south pole of the earth. The magnetic south pole of the

geographic north pole.

concentric circles surrounding the wire. Each of these circles has its centre at the wire.

dutifully follow it.

Try It Yourself

Hint:

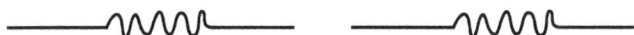

Project-19

Hand Battery

Do you know your skin and two different metals create a battery?

Introduction

An electric battery is a device that converts stored chemical energy into electrical energy. It contains a positive electrode and a negative electrode, that conduct electricity, and a third substance, an electrolyte, that acts chemically on the electrodes. The electrolyte functions as an ionic conductor for the transfer of the electrons between electrodes. You can make a battery with your hands. When you place your hands on metal plates, you and the plates form a battery.

Materials Required

- A DC micro ammeter (capable of reading 100 micro amps)
- An aluminum plate and a copper plate (each about the size of your hand)
- 2 electrical lead wires with alligator clips at both ends (available at radio shock)
- **Optional–** FET input voltmeter that can measure upto 2 volts; a piece of wood or non-metallic surface; plates made of other metals such as lead or zinc (galvanized steel).

Assembly

Mount both metal plates on a piece of wood or simply clamp on a non-metallic surface.

Using the clip leads, connect one plate to one of the meter's terminals and other plate to the other terminal. At this point it does not matter which plate attaches to which terminal.

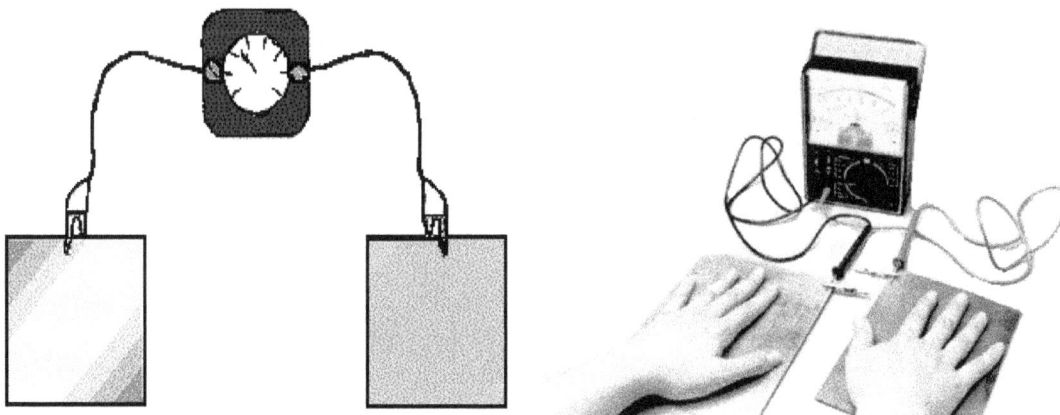

Fig. 19.1 : Circuit Diagram

To Do and Notice

Place one hand on each plate. You should notice a reading on the meter. If the meter does not

connections and the wiring. If that does not produce current, try cleaning the plates with a pencil

What Happens?

battery, reacting with the copper plate and with the alumnium plate. In one of these reactions,

meter will show a small current.

Electrons pile up on the aluminum side because they can not cross the gap in the wire. The copper

Try It Yourself

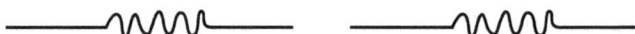

Project-20

Static Electricity

How do you reveal the electrifying nature (+ve or -ve) of an object with respect to glass or plastic rod?

Introduction

Static electricity is an imbalance of electric charges within or on the surface of a material. Metarials are made of atoms that are normally neutral because they contain equal number of +ve and –ve charges. When two materials are in contact, electrons may move from one material to the other, which leaves an excess of positive charge on one material and an equal negative charge on the other. When materials are separated, they retain this charge in imbalanced way. This can be seen by stroking glass and plastic road, by woollen cloth: the glass rod charged +vely and the plastic rod charged -vely equal to that of +ve charge of glass.

Materials Rrequired

- A glass jar
- A cork which fits tightly enough to act as a seal
- A length of iron wire
- A strip of tinfoil
- A glass rod and a plastic rod
- A woollen cloth

Assembly

Note : Once the experiment has been set up, take care not to touch the iron wire with your hand, otherwise the electrical charge will be lost.

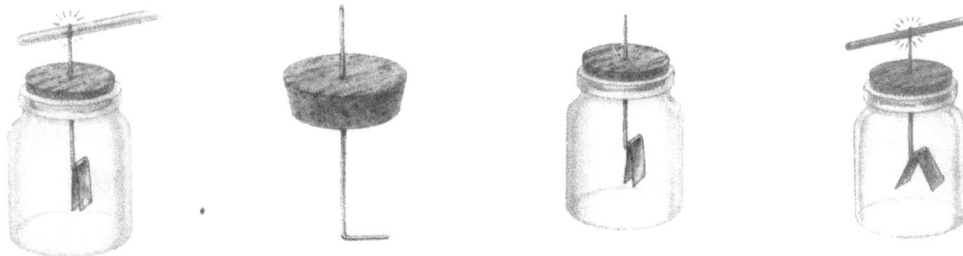

1. Thread the wire through the centre of the cork so that it sticks out at the top and at the bottom. Bend the bottom end as shown in the picture.

plastic rod.

To Do and Notice

woollen cloth. Touch the top wire with the glass rod.

What Happens?

Try It Yourself

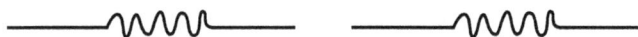

Single Pole, Double Throw Switch or 2-Way Switch

How do you control ON/OFF action of a lamp from two different places?

Introduction

While wiring a building, the multiway switching is the interconnection of two or more electrical switches to control an electrical load from more than one location. While a normal light switch needs to be only single-pole single-throw switch, the multiway switching requires the use of switches that have one or more additional contacts and two or more wires must be run beetween the switches. When the load is controlled from only two points, single-pole double-throw (SPDT) switches are used.

Materials Required

- Electric cable
- 6 drawing pins
- Two paper clips
- A 4.5V battery
- Two small tablets of wood
- A bulb and a bulb holder

Assembly

1. Push three drawing pins into each tablet of wood, as you see in the picture.

2. On each wooden tablet, open up a paper-clip and place one end under the centre drawing pin. In this way, the paper-clip can be moved to touch either of the other two drawing pins.

3. Using the electric cable, connect switches with the battery and the bulb, as shown in the fig.

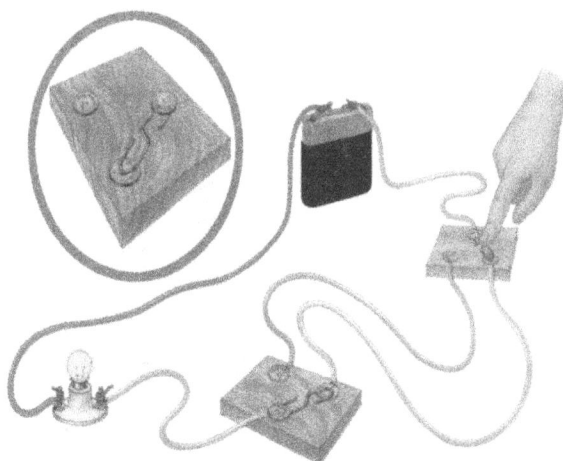

Fig. 21.1 : Circuit Diagram

near to each door. In this case, an electrical installation must be equipped with a two-way switches.

staircase.

To Do and Notice
The bulb can be lit or extinguished by either of the two switches.

What Happens?

makes the light go out.

Try It Yourself

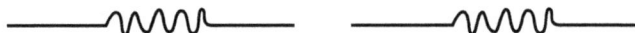

Project-22

Automatic Water Pump Cut Off

Can you make an electronic device which controls the ON and OFF timings for a fixed duration?

Introduction

In multistoried complexes several flats are provided for residential and office purposes. In many cases, each individual house or flat has its own water tank on the top roof, sometimes several storeys high. Since water storage is a perennial problem, wasting any amount of water by overflowing storage tanks is unpardonable. The problem lies in accessibility water tank for personal inspection and monitoring. An easy solution to the problem is to have an electronic device to operate the pump for a fixed duration and then cut it off.

Materials Required

- Transformer (T_1), bridge rectifier (B.)
- Filter capacitor (C_1), AC power supply switches (S_1, S_2)
- Relay switch
- LED
- Field-effect transistor
- Resistors (R_1, R_2, R_3)
- Diodes (1, 2)
- Capacitors (C_1, C_2, C_3) and connecting wires

Assembly

In the fig. 22.1 the circuit transformer T_1, bridge rectifier B and filter capacitor C_1 form the power supply for the electronic timer circuit. Switch S_1 controls power to the whole unit as well as the pump itself. Switch S_2 provides the option to operate the pump through the timer or by-pass the time and run the pump directly through the AC mains.

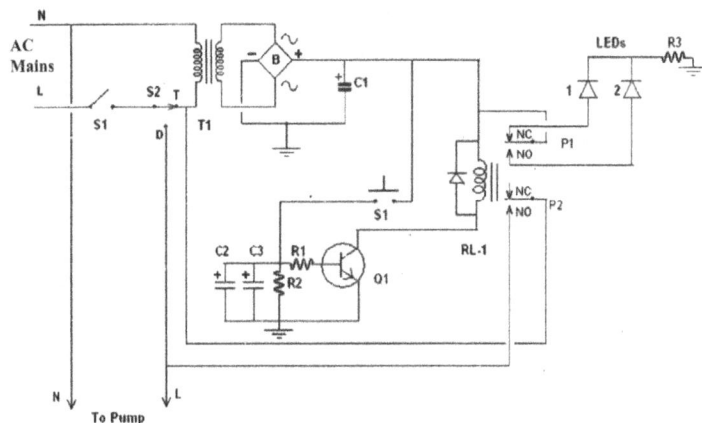

Fig. 22.1 : Circuit Diagram

To Do and Notice

Put the switch S $_1$

$_1$ conducts and the

relay is switched-on.

.

What happens?

discharge slowly through R

and the relay deeps back into its original position.

and R $_1$ is maintained for about thirty
pair.

Similarly, the time period for operating the pump can be decreased by decreasing the total
$_1$, S and the

relay contacts, hence they should be rotated to carry the operating current for the pump.

Try It Yourself

Make an automatic water pump   system for underground water tanks.

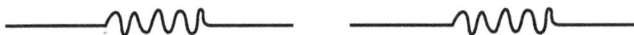

Project-23

Electronic Switch Function

What do you know about working and assembly of electronic switches?

Introduction

An electronic switch is an electronic component or device that can switch an electrical circuit, interrupting the current or diverting it from one conductor to another. Typically, electronic switches use solid state devices such as transistors though vacuum tubes can be used as well in high voltage applications. In this experiment we learn the operation of electronic switches using electronic devices like transistors or SCRs. We have chosen these two as they are the most commonly used ones in this role, and in the electronic circuits of hobbyist's' interrerst.

Materials Required

- **Resistors** : All resistors of carbon film or carbon composition type
  $R_1, R_2, R_3, R_4, R_5, R_6$: 22K, 1/4 W
  R_9, R_{10} : 3.3 K, 1/4W
  R_7 : 470Ω, 1/4W
  R_8 : 470Ω, 1/4W
- **Potentiometers**
  P_1 : 10K preset
- **Transistors**
  Q_1, Q_3 : 2N2222
  Q_2 : 2N2907
- **Diodes**
  D_1 : 1N4001
- **SCR**
  SCR-1 : SN100 or any other 200 volts SCR
  LED-1 to LED-4 : Red LEDs
- **Relay**
  RL-1 : 6 V DC relay
- **Miscellaneous**
  Battery : 9 V DC battery
  Switches SW_1 to SW_5 : ON/OFF switches

Assembly

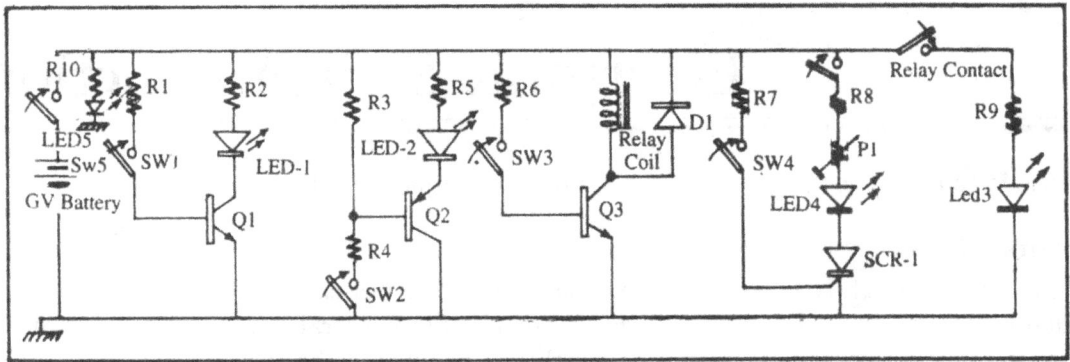

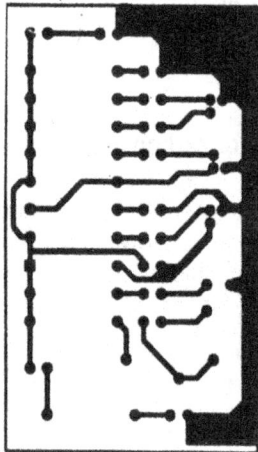
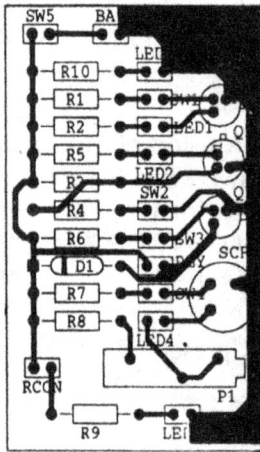
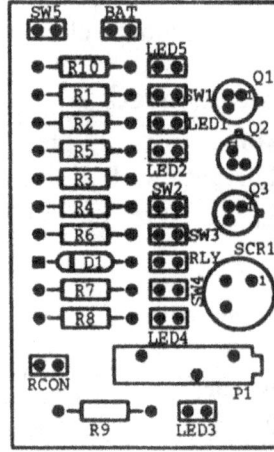

$_1$, R

switch.

The second part consisting of resistors R , R , R   , depicts the
use of a PNP and transistor as a switch.

(as the case may be) closed in order to keep the transistor

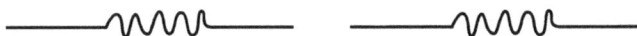

To Do and Notice

$_1$ $_1$ is closed.

                                                  $_1$

                                              $_1$ is opened.

What Happens?

                                                                    . There is only one method to bring

Try It Yourself

————⌁⌁⌁————   ————⌁⌁⌁————

Project-24

Car Battery Charger

How will you charge your car battery in the event of misbehaving or having become weak, what do you require in addition to 220V AC mains?

Introduction

A battery charger or recharger is a device used to put energy into a rechargeable battery by forcing an electric current through it. The charging protocol depends on the size and type of the battery being charged. Here is a simple battery charger circuit, the one that can be used to charge 12 V batteries of both the usual automobile type as well as the maintenance free sealed lead-acid type. The charger circuit described here is quite compact and can be placed right on the top of the battery required to be charged.

Material Required

- Resistors and capacitors
  R_1 :   2.2 K, 1/4W
  C_1 :   1000 µF, 25V (electrolytic)
  C_2, C_3 :   0.1 µF (ceramic disc)
- Semiconductor devices and ICs
  D_1, D_2, D_5 :   BY 127
  D_3, D_4 :   1N4001 or equivalent
  IC-1   :   78T12 (it is 7812 in TO-3 package)
  LED-1   :   LED (any colour)
- Miscellaneous
  Meter M-1   :   Ammeter 0-3A (fig. 10.4)
  M-2   :   DC Voltmeter 0-1V DC
  Transformer T-1 :   15-O-15, 2A
  Fuse F-1   :   1A tubular type with holder

Power supply terminals, solder metal, wires, main power ON/OFF switch.

Assembly

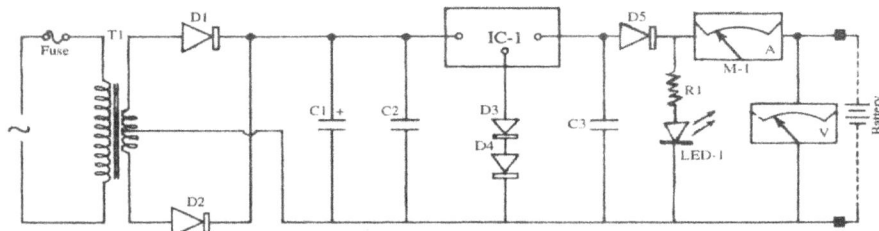

Fig. 24.1 : Circuit Diagram

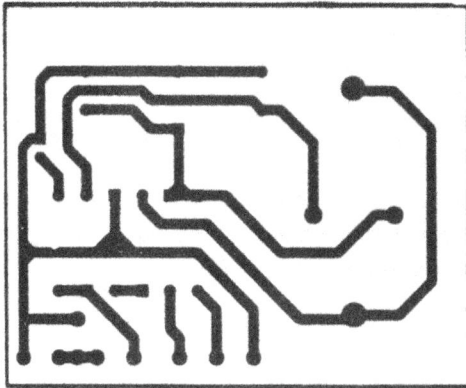

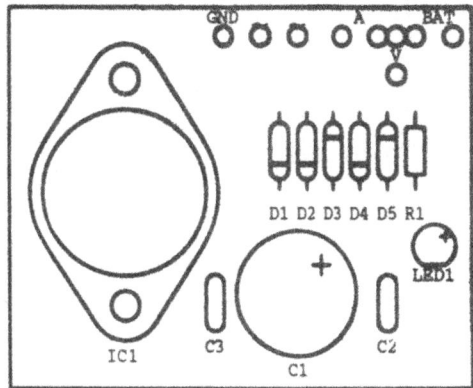

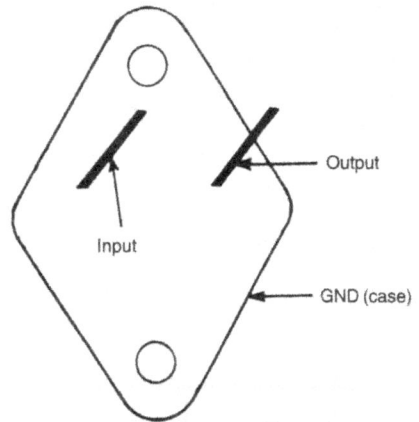

and

$_1$, consititute the unregulated power supply

$_1$

$_1$

common terminal of this regulator has been lifted to a

are decoupling

To Do and Notice

resistance across the output in place of the battery.

What happens?

Try It Yourself

Project-25

Multi Purpose Power Supply

Can you make regulated and put DC voltages of 3V, 6V, 9V and 12V that are seleted one at time with the help of a rotary switch, from a single AC (220-230) source?

Introduction

A regulated power supply is one that controls the output voltage or current to a specific value despite variations in either load current or voltage supplied by the power supply's energy source. To produce DC from an AC powered supply, a rectifier is used to convert AC voltage to a DC voltage. Here we intend to build a multi-purpose regulated power supply that will serve as a very useful alternative to our multiple needs of different battery eliminators to operate gadgets like transistor sets, Audio Casette recorders and so on and so forth.

Materials required

- **Resistors and capacitors**

R_1	:	3.9 ohms, 2 watt
C_1	:	1000 µF, 25V (electrolytic)
C_2, C_3, C_4	:	0.1 µF (ceramic disc)

- **Semiconductor devices and ICs**

Diodes D_1 to D_4	:	1N4001 or equivalent
Zener diode V_{Z1}	:	3V, 2 watt
IC_1	:	7812
IC_2	:	7809
IC_3	:	7806

- **Miscellaneous**

Transformer T_1	:	0-4, 5-7.5-12-15, 500 mA mains transformer
Fuse F_1	:	Tubular type 500 mA fuse with holder
SW_2	:	Main power ON/OFF switch
SW_1	:	3-pole, 4- throw rotatory switch

Power supply terminals, solder wire, mains cord etc.

Note : In case IC_2 and IC_3 are not easily available) IC_2 (7809) can be replaced by a 7805 with two series connected silicon diodes of 1N4001 type wired between the IC's common terminal and circuit GND.

IC_3 (7809) can also be replaced by a 7805 with a 3.9 zener diode wired from the IC's common terminal to circuit GND.

Assembly

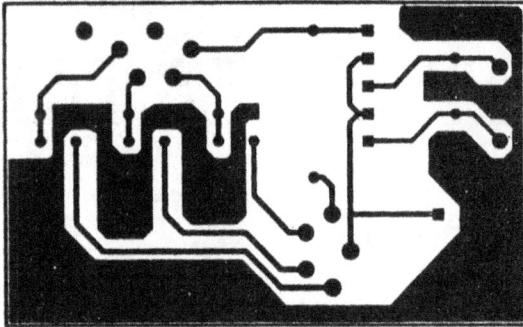

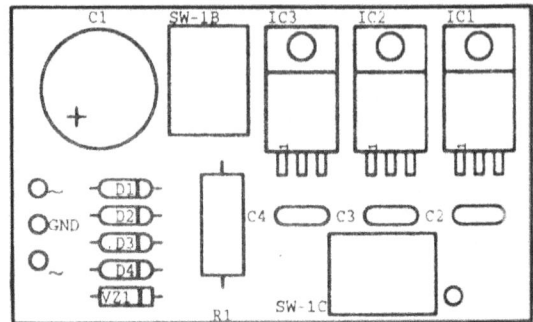

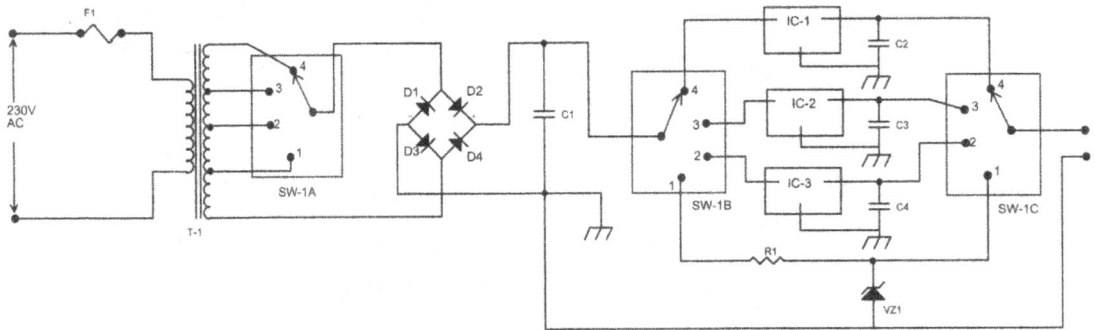

The transformer T-1 is a mains transformer with a multiple secondary winding that has taps at

are decoupling capacitors.

To Do and Notice

What Happens?

Try It Yourself

Project-26

Fully Automatic Emergency Light

Can you build a fully automatic emergency light that remains OFF in the normal circumstances and switching ON automatically in the event of mains power failure?

Introduction

An emergency light is a battery-backed lighting device that comes on automatically when a building experiences a power outage. Such gadgets are very useful when used in places such as electricity meter box or the place where various cut-outs are located. The other useful places for these gadgets are dark corners where a little bit of light in case of power failure is of great utility.

Material Required

- **Resistors and capacitors**

R_1	:	27Ω, 1/2 W
R_2	:	470Ω, 1/4 W
C_1, C_3	:	0.1 µF, 25V (ceramic disc)
C_2	:	1000 µF, 16V (electrolytic)

- **Semiconductor devices and ICs**

D_1, D_2, D_3	:	1N4001 or equivalent
Q_1	:	2N2907
VR-1	:	Three terminal regulator, type no. 7809

- **Other components**

S_1	:	Main power ON/OFF switch
F_1	:	Fuse (0.5 A rating) with holder
T_1	:	Main transformer
		Primary : 230V AC
		Secondary : 12-0-12. 250 mA
Lamp	:	6V DC lamp
Battery	:	5 × 1.2 V, Ni-Cd cells

Assembly

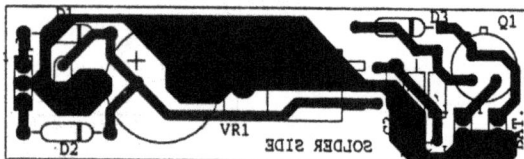

Fig. 26.1 : Circuit Diagram

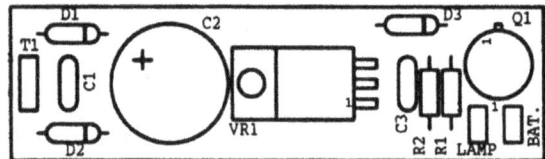

Fig. 26.2 : Circuit Diagram

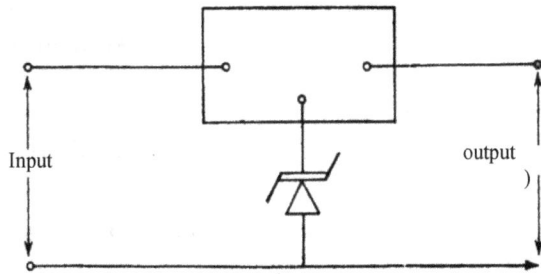

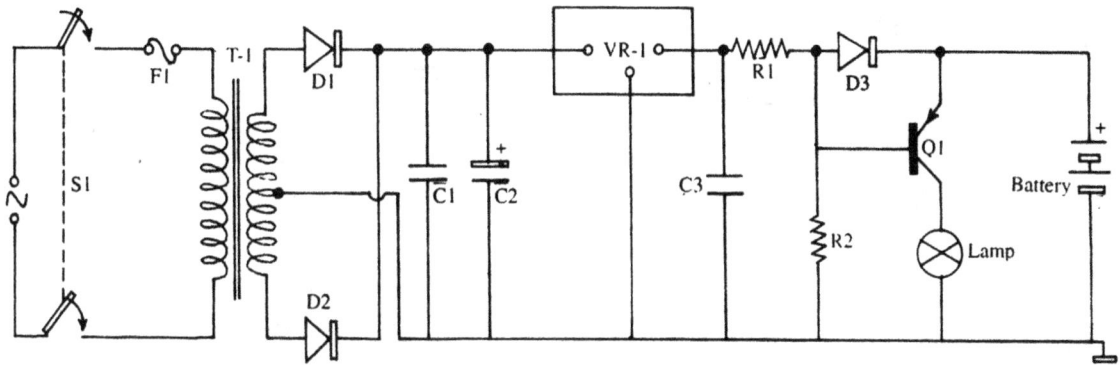

to ground instead of grounding the common terminal.

is the step

To Do and Notice

ensures that the emitter potential of transistor is less than that of its base potential.

$_1$) remains in cut-off. R_1 is so chosen that the charging current for the Ni-

$_1$) is

) is forward biased. The lamp is lighted.

What Happens?

$_1$ remains in cut-off and the battery is charged

$_1$ conducts. The lamp is lighted.

The lamp extinguishes automatically when the mains power is restored. The charge drained out of

Try It Yourself

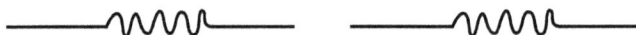

Project-27

Staircase Light with Auto-Switch-OFF Feature

Can you wire an electrical bulb with two switches with one of the switches placed at the bottom of a staircase and the other located at top?

Introduction

We are all familiar with the electrical wiring diagram. The wiring is done is such a way that either of the two switches can be used to switch ON the bulb (if it is initially OFF) or switch it OFF (if it is initially ON).

In the present circuit, we have two micro-switch with every push of either of the two switches the bulb lights up for a time period of about 40 seconds after that the light will turn automatically OFF.

Materials Required

- **Resistors**
  R_1, R_2, R_3, R_4 : 22K, 1/4W
  R_5, R_6 : 47K, 1/4W
  R_7 : 12K, 1/4W
  R_8 : 3.9K, 1/4W
  R_9 : 1.5K, 1/4W
  R_{10} : 4.70Ω, 1/4W
- **Capacitors**
  C_1, C_2, C_4 : 0.1 μF, 25V (ceramic disc)
  C_3, C_6 : 0.01 μF (ceramic disc)
  C_5 : 10 μF (Tantalum)
- **Semiconductor devices and ICs**
  Diodes D_1, D_2, D_3 : 1N4001
  Transistor Q_1 : 2N2222
  Voltage regulator, VR_1 : 7805
  IC_1, IC_2 : CD4011
  IC_3 : timer 555
- **Hardware components**
  SW_1, SW_2 : Micro-switches
  SW_3 : ON/OFF toggle switch

relay with at least one normally

Assembly

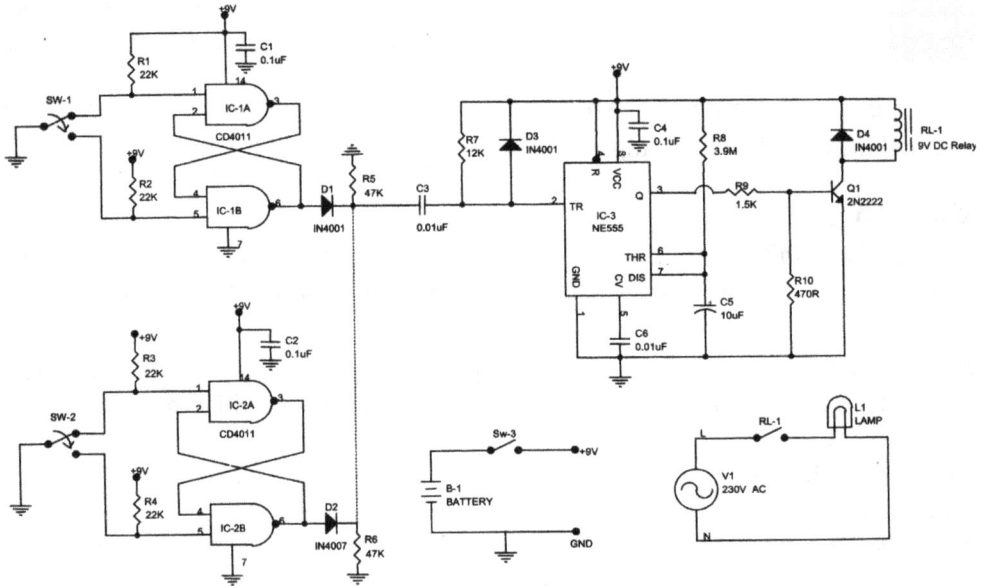

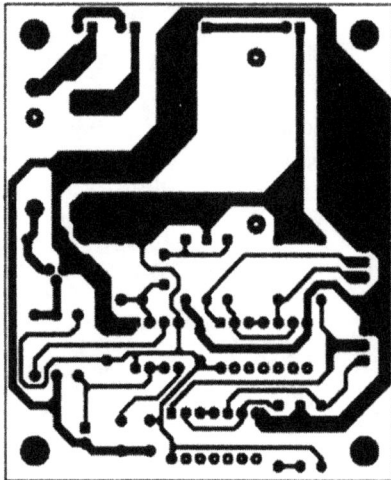

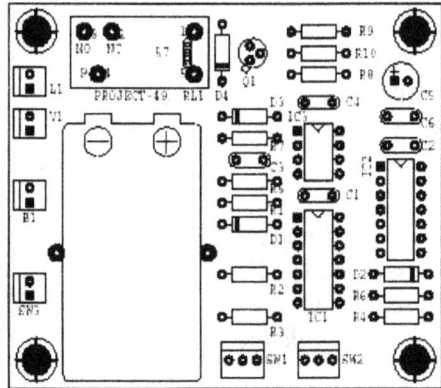

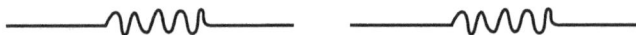

are the two micro-switches, which feed the

To Do and Notice

. This pulse is

contact wired in series with the mains and the bulb.

What Happens?

Try It Yourself
Make an electrical arrangement to get a similar facility.

Project-28

Sound Operated Light

Can you switch ON the lights with the sound of a clap?

Introduction
This gadget has a sensitivity adjust control so that the gadget can be made to operate at the intended sound level. Gadget getting triggered by a whisper and the gadget not responding to even a lound bang, both are undesirable situations.

The proposed gadget is designed to switch on and off alternately. That is if one sound pulse switches the light on, another one will switch it off. It is a desirable feature when you install this gadget with the sole aim of switching the lights ON and OFF at will without having to go to the switch board. Even if the intension is to keep the burglars away, alternate switching of lights could really frighten them away.

Materials Required
- **Resistors and capacitors**

R_1	:	10K, 1/4W
R_2	:	1K, 1/4W
R_3	:	470Ω, 1/4W
R_4	:	2.2K, 1/4W
P_1	:	1K preset
C_1	:	1000 μF, 50V (electrolytic)

- **Semiconductor**

D_1, D_2	:	1N4001 or equivalent
Q_1	:	2N2222
IC_1	:	741
IC_2	:	CD4027B

- **Components**

Transformer T_1	:	12V, 250 mA mains transformer condenser microphone
Relay	:	12V DC relay with at least one normally open contact
Fuse	:	0.5A with holder
Switch	:	S_1 (Main ON/OFF switch)

- **Miscellaneous**

  Solder metal, multistrand wires, suitable cobinet etc.

Assembly

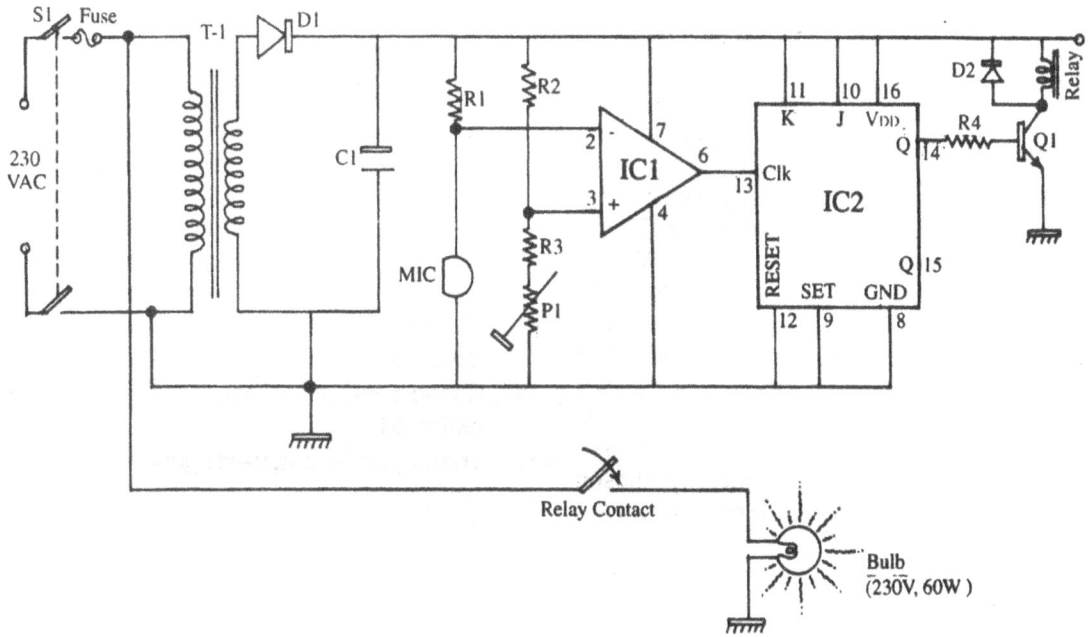

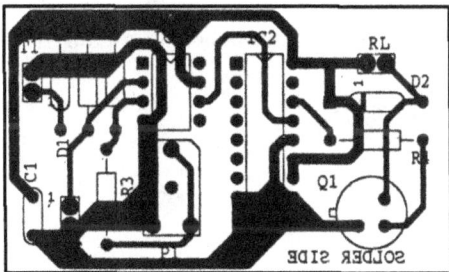

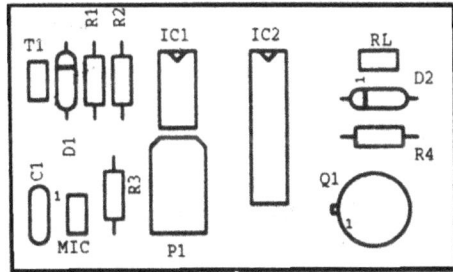

) whose ‾

₁) wired

Transformer T₁   ₁   ₁

produced by R , R and the preset P₁. P₁

To Do and Notice

out to establish the gadget reliability and that it does not trigger on false alarms.

What Happens?

$_1$) is in cut-off and the relay remains de-energised.

remains extinghished.

1

toggles

opens. The bulb extinguishes again. Thus the bulb lights up and extinguishes alternately if there are re-current sound pulses.

Try It Yourself

of a door.

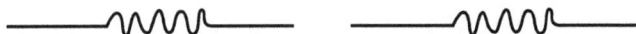

Project-29

Digital Dice

Can you display in random the number from 1 to 9 on the 7 segment display?

Introduction

The digital dice project is an interesting project that will display in random the number 1 to 9 on the 7 segment display. This is an alternative device that can be used to replace the traditional dice when you are playing games such as Snake Ladder and Monopoly. This circuit is mostly for exercising and fun.

Materials Required

U_1	:	7490 decade counter
U_2	:	7483 4-bit binary adder
U_3	:	7447 BCD to 7 segment decoder
U_4	:	Common anode 7 segment display
U_5	:	555 timer
R_1, R_2	:	1kΩ, 1/4W, 1% resistor
R_3, R_9	:	270Ω, 1/4W, 5% resistor
PB	:	Normally open push button
C_1, C_2	:	0.01 µF/25V ceramic capacitor
E_1	:	220 µF/16V electroyltic capacitor

Assembly

The generation of clock is done by using a 555 timer which is connected in the astable mode at a frequency of approximately 50 Hz. The clock signal is fed into the decade counter which outputs are connected to 4 bit binary adder which provides a binary output equivalent to binary input +1. The outputs are then connected to a BCD to 7 segment decoder which is used to drive a common anode 7 segment display.

Fig 29.1 : Circuit Diagram

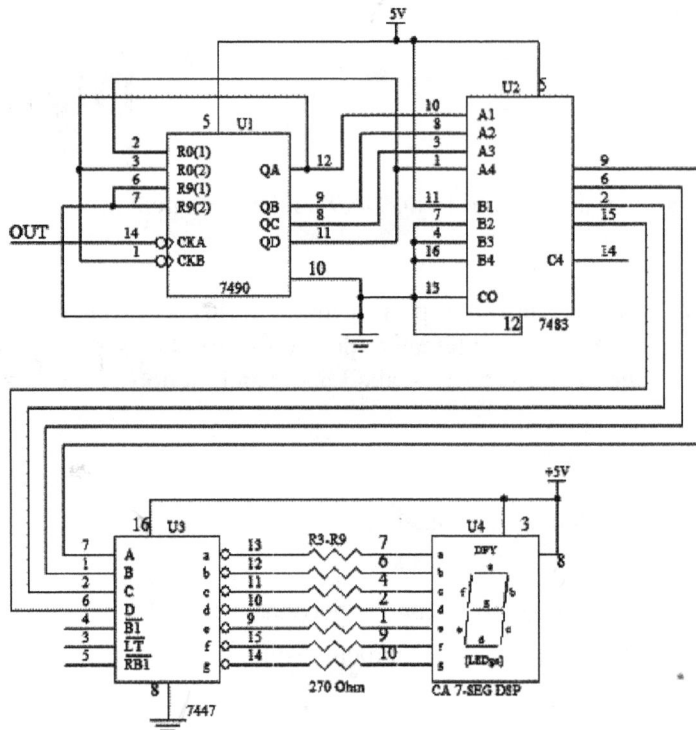

To Do and Notice

What Happens?

1

Try It Yourself

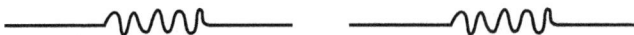

Project-30

Battery Tester

Can you design and build a battery tester to test dry cell and rechargeable battery with a voltage less than 2V?

Introduction

The objective of this project is to design a battery tester using LM 3914 IC that is able to test various types of dry cell and rechargeable battery with a voltage of less than 2V. The LM 3914 IC from national semiconductar senses the voltage levels of the battery under test and drives the 10 LEDs to ON or OFF based on the voltage that is detected. The current driving the LEDs is regulated by using the enternal resistor R_1 and hence limiting resistors are not required.

Materials Required

U_1	:	LM3914 IC
L_1, L_2, L_3, L_4, L_5, L_6, L_7	:	5 mm Red LED
L_8, L_9, L_{10}	:	5 mm Green LED
S_1	:	SPST toggle switch
R_1	:	1KΩ, 1/4W, 1% resistor
R_2	:	100Ω, 1/4W, 1% resistor
VR_1	:	1KΩ variable resistor
C_1	:	0.1 µF/25V ceramic capacitor
T_1, T_2	:	Probes

Assembly

The schematic shows the simple connections where the reference voltage at pin 8 of U_1 can be adjusted by adjusting the variable resistor VR_1. The voltage at pin 8 will set the maximum scale of the LED. In testing dry cell battery of 1.5V, set the voltage at pin 8 to 2.0V. Each of the LED will thus represent 200 mV when lighted up.

Test the circuit by using a breadboard

This is to ensure that all the components that are used are in working condition.

To Do and Notice

Testing of reachargeable battery such

a rechargeable battery is approximately

the polarity of the probe to the terminals of the battery. T_1 is to be placed on the

terminal of the battery.

What Happens?

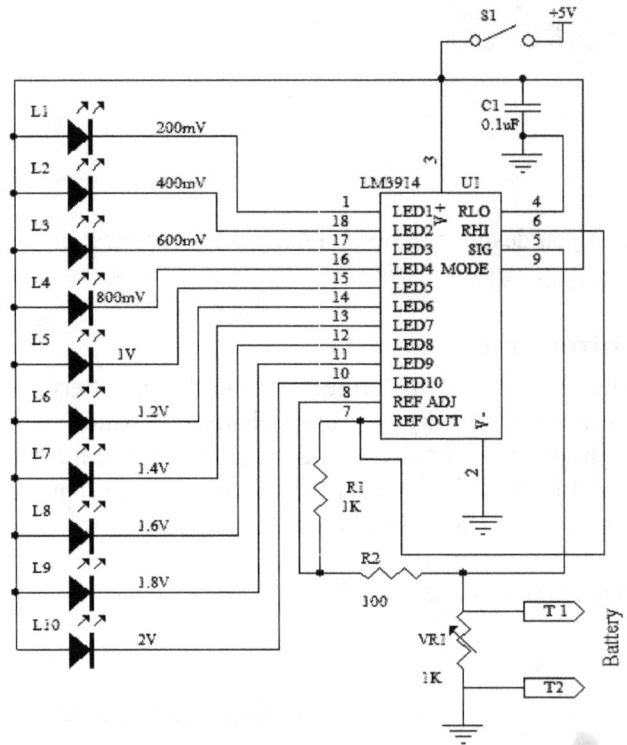

Try It Yourself

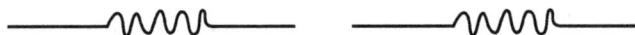

Project-31

Home Security alarm

Can you build a security system for your home?

Introduction

A home alarm is a system designed to detect intrusion – unauthorized entry into a building. This simple stand alone home alarm project is a good project for those who are eager to know the concept of home security system. This circuit which consists of three 555 timers gets the input from a contact that could be connected to a motion detector or any momentary contact that monitors a certain process.

Materials Required

U_1, U_2, U_3	:	555 timer IC
R_1, R_2, R_4, R_6	:	10KΩ 1/4W, 5% carbon film resistors
R_3	:	82KΩ 1/4W, 5% carbon film resistor
R_5	:	200Ω 1/4W, 5% carbon film resistor
VR_1	:	1MΩ potentiometer
VR_2	:	200KΩ potentiometer
C_1, C_2, C_3	:	0.01 µF/25V ceramic capacitor
E_3	:	10 µF/25V electrolytic capacitor
E_1, E_2	:	220 µF/25V electrolytic capacitor
D_1, D_2	:	Diode 1N4003
SPKR (Speaker)	:	8Ω speaker
Q_1	:	Transistor TIP41A
S_1	:	SPST switch
CT	:	Dry contact to trigger the alarm.

Assembly

The circuit diagram of the project is shown in fig 31.2.

The first timer U_1 is configured as a monostable timer with a variable timing up to a maximum of 220 seconds. The timing can be changed by ajdusting the VR_1 1MΩ potentiometer. If you need to further increase the timing, the electrolytic capacitor E_1 can be changed to a larger value.

and
timers causing both timers to start their astable

, R and R .

) (E

timers is used to feed control
timer through register R_6 where

frequency modulation will generate a tone similar
to the police siren.

To Do and Notice

is

₁ due to the back emf generated by the speaker during

.

.

What Happens?

second timer is then fed into the third timer to control its frequency modulation. This frequency modulation will then power a speaker and generate a siren tone similar to the police siren.

Try It Yourself

Make a simple laser based intruder alarm.

Project-32

LED Light Flasher

Can you build LED Light Flasher with two LEDs connected in parallel?

Introduction

This is a simple LED light flasher project that uses a CMOS 74CO4 IC to alternately ON and OFF two LEDs that are connected in parallel. The hex inverter MM 74CO4 from Fairchild Semiconductor has a wide operating power supply voltage range from 3V to 15V DC. It has a typical low power consumption of 10μW/package and has high noise immunity. It is back to back compatible with the standard 74 logic family which is freely available in the market. All its inputs have diode clamps to V_{cc} and GND which protect them from damage due to electrostatic discharge.

Materials Required

R_1, R_2, R_3	:	10MΩ resistor 1/4W 5%
R_4	:	510Ω resistor 1/4W 5%
C_1	:	0.1 μF/25V ceramic capacitor
S_1	:	SPDT switch
IC_1	:	MM 74CO4 Hex inverter
LED_1, LED_2	:	3 mm or 5 mm LEDs
9V	:	9V battery and battery clip

Assembly

The schematic above shows the simple configuration of the project. It uses two inverters U1A and U1B form an oscillator configuration where the frequency of the oscillation is given by

$$f \quad = \quad 1/[1.4 \; RC]$$
$$= \quad 1/[1.4 \; (10M\Omega) \; (0.1\mu f)]$$
$$= \quad 0.7 \; H_z.$$

The square wave frequency of 0.7 H_z. is used to feed the input of U1D which is used as a buffer circuit. At the same time, the other inverter U1C gets its input from Pin 2 and Pin 3 of V1. With this configuration, when U1D output is high, U1C output will be low and vice-versa. In this way when LED_1 is ON, LED_2 will be off and this will alternate at a frequency of 0.7 H_z. The current that goes through LED is given by

$$I \quad = \quad (9V - 7V/510 \; \Omega)$$
$$= \quad 14 \; \mu A$$

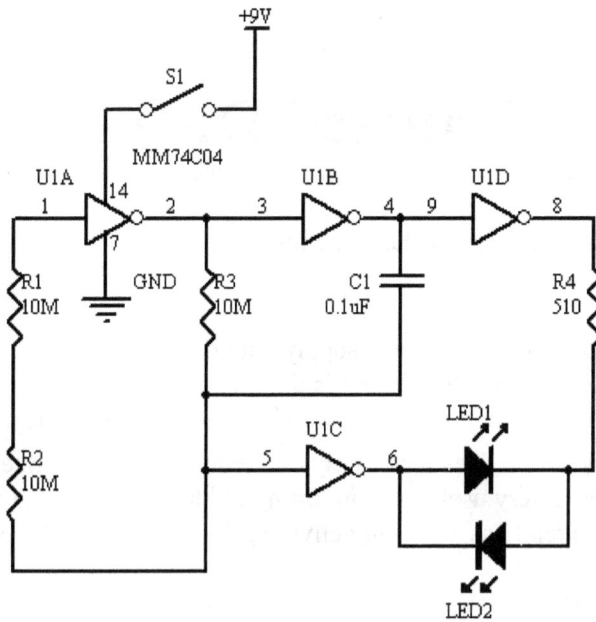

To Do and Notice

[1]

What Happens?

[1], R , R [1]. The brightness of the

Try It Yourself

Project-33

Regulated Dual Power Supply

Can you build a dual power supply that generates +12 V and –12 V from 220-230 V AC mains?

Introduction

Some electronic circuits require a power supply with positive and negative outputs as well as zero volts. This is called dual supply because it is like two ordinary supplies connected together. Such a supply is a very common requirement in all those circuits that use opams.

Since opams are very widely used in a variety of circuits of hobbyists interests, construction of this project could serve as a very useful tool in testing all those circuits that need a dual supply. Each of the outputs in the circuit has a current delivering capability of 250 mA.

Materials Required

C_1, C_2	:	1000 μF, 50V electrolytic capacitor
C_3, C_4, C_7, C_8	:	0.1 μF ceramic disc capacitor
C_5, C_6	:	10 μF, 50V electrolytic capacitor
D_1 to D_4	:	1N4001 or equivalent diodes
IC-1	:	7812
IC-2	:	7912
SW-1	:	Mains ON/OFF switch
T-1	:	150-0-15V, 250 mA mains transformer
F-1	:	500 mA tubular type with holder fuse

Assembly

The unregualted AC/DC power supply part of the circuit consist of a transformer (T-1) that steps down 230V AC to 15V across a centre tapped secondary winding. Diodes D_1 to D_4 that rectify AC appearing across the secondary with D_1 and D_3 providing full wave rectification to produce positive output, D_2 and D_4 providing full wave rectification to produce a negative output, capacitor C_1 and C_2 providing the filtering action. C_3 to C_8 are decoupling capacitors. IC-1 is a fixed output positive three terminal regulator where IC-2 is a fixed output negative three terminal regulator.

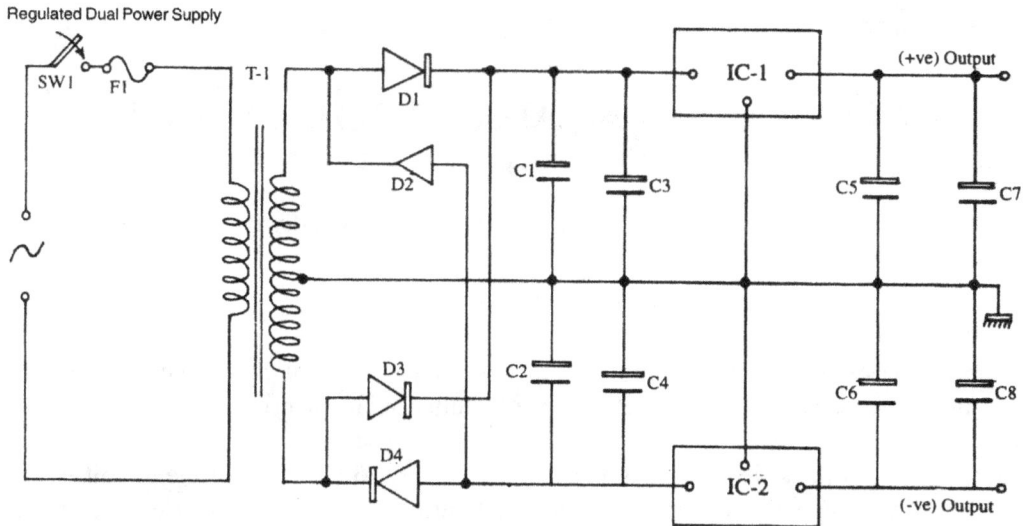

Regulated Dual Power Supply

To Do and Notice

a multimeter.

What Happens?

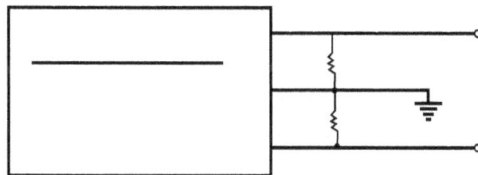

Try It Yourself

Project-34

Automatic Night Light

Can you make simple circuit which make a light bulb automatically on when the day light fades into a dark night?

Introduction

This project needs an extremely simple circuit consisting of just four essential components to build a fully automatic self sensing night light. The light bulb used in the project could be the porch light that would go ON as the day light fades into a dark night. The same could also be used in the open terrace. There is a provision for disconnecting AC supply to the input of the circuit in case you deliberately want the lamp to remain extinguished during night due to some reasons or the other. One such instance could be when you are going out of the house during the daytime and you are not likely to return the same day.

Materials Required

R	:	10M Ω, 212W carbon film resistor
LDR	:	
SCR	:	Any 2A 600 V SCR such as OE 206, SN 206, 2N2601, TY6004
Triac	:	Any 2A, 600V Triac such as SPT 6M, 2N5757
Lamp	:	60W lamp
S-1	:	Mains ON/OFF switch
S-2	:	DPDT switch
Fuse	:	3A rating

Assembly

The circuit description is straight forward once operational principles of SCR and Triac are clearly understood. We know that an SCR is a three terminal device with anode, cathode and gate as its terminals. It behaves as an open switch as long as the anode – cathode voltage is less than the break over voltage of the device. An SCR remains in the open state when anode is negative with respect to cathode. Triac behaves like a combination of SCRs connected in inverse parallel with the result that for both positive as well as negative anode – cathode voltages, the triac can be made to switch to the ON state by applying an appropriate gate signal. Anode and cathode of an SCR are designated as MT – 1 (main terminal – 1) and MT – 2 (main terminal – 2) in case of Triac.

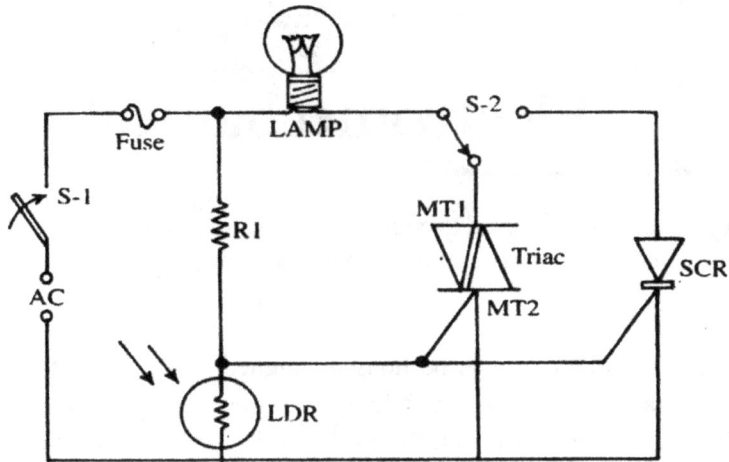

To Do and Notice

What Happens?

Try It Yourself

Project-35

Electromechanical Counter

Can you build a project which can be used to count the number of bottles passing over the conveyer belt or even the number of people passing through a certain entrance?

Introduction

In this project, the transmitter and receiver portions are aligned with each other and infra-red light beam generated by the transmitter is directly falling on the photo diode in the receiver. Whenever the infra-red light beam is interrupted, the electromechanical counter beam advances by one count. In a typical application of counting of bottles passing over a moving conveyer belt, the transmitter and receiver portions are located on the two opposite sides of the belt and aligned with each other to establish a continuous light bream path. The light path is interrupted every time a bottle moves across it. The number of bottles passing over a period of time is thus known from counter reading.

Materials Required

R_1	:	68Ω, 1W resistor
R_2	:	10MΩ, 1/4W resistor
R_3	:	1KΩ, 1/4W resistor
R_4	:	470Ω, 1/4W resistor
R_5	:	4.7Ω, 1/4W resistor
D_1	:	1N4001 or equivalent diode
D_2, D_3	:	Infra-red emitting diode type TIL34
D_4	:	Photo diode, type TIL81
D_5	:	LED
Q_1, Q_2	:	2N2222
IC-1	:	LM324 (quad opamp)
S_1, S_2	:	two ON/OFF switches

Two 9 V batteries, and electromechanical incremental counters (14 pin DIL IC socket)

Assembly

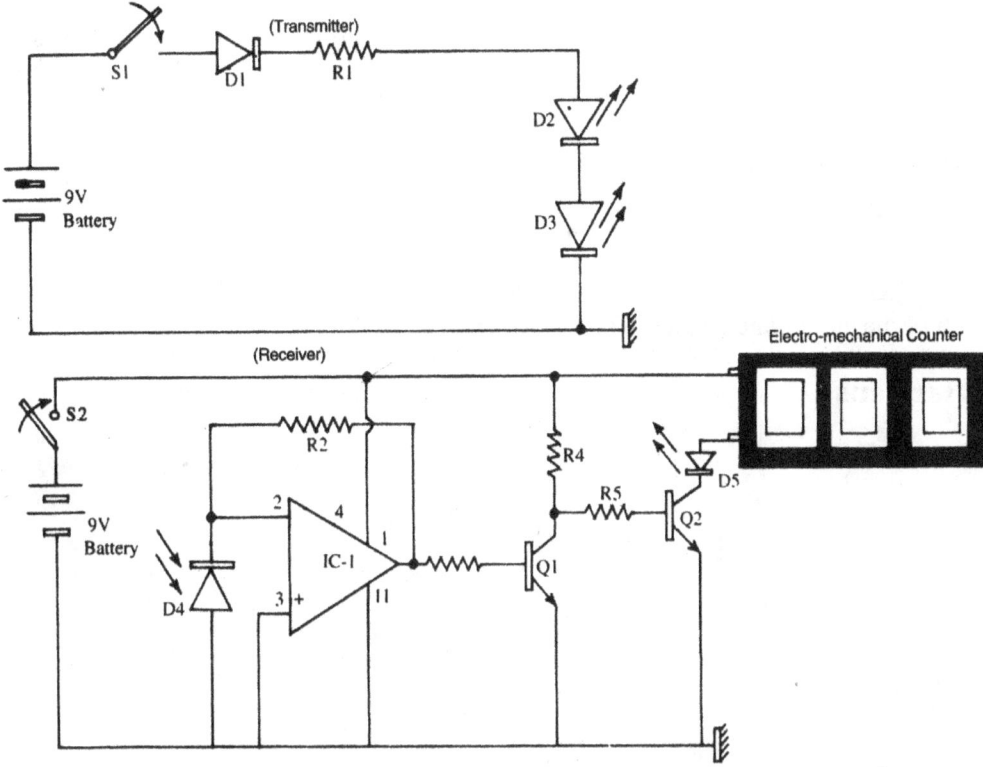

(Transmitter)

(Receiver)

Electro-mechanical Counter

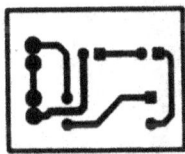

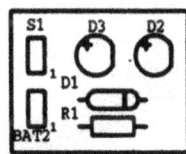

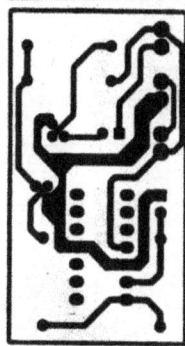

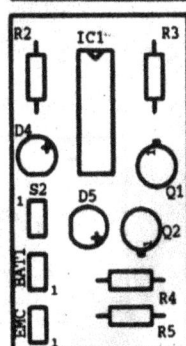

once aligned with the transmittter module senses any interruption of the light beam caused due

counter by one count.

To Do and Notice

count. If the light beam remains blocked continuously which does not happen in the intended application of the circuit, the counter would keep on incrementing at a steady rate.

What Happens?

$_1$. The current has been

in the off-state with the

result that the counter stays its initial setting.

R   and R

Try It Yourself

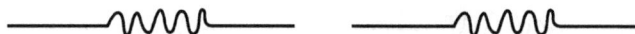

Project-36

Invisible Intruder Alarm

Can you make an intruder alarm for a locker which has audio as well as visible indicators?

Introduction

This is an ultra compact intruder alarm system based on the detection of an intrusion caused by the interruption of an infra-red light beam. This gadget has transmitter and receiver portions separately which is operated separately by 9V batteries. The transmitter and receiver gadgets can be mounted in an aligned position on the two facing walls of inside of a locker to give you audio (micro buzzer) and visible (LED) indication alarm. The LED can remotely be located away from the location of gadget also.

Materials Required

R_1	:	68Ω, 1W resistor
R_2	:	10MΩ, 1W resistor
R_3	:	1KΩ, 1/4W resistor
R_4	:	470Ω, 1/4W resistor
R_5, R_6	:	4.7KΩ, 1/4W resistor
R_7	:	100Ω, 1/4W resistor
D_1, D_6	:	IN4001 or equivalent diode
D_4	:	Photo diode, type No. TIL81
D_2, D_3	:	Infra-red emitting diode type No. TIL34
D_4	:	Infra-red emitting diode type No. TIL34
D_5, D_7	:	LEDs (D_7 red and D_5 green)
Q_1, Q_2	:	2N2222 transistors
SCR	:	OE101 or equivalent (any SCR with 100V breakdown voltage and 1A forward current rarings)
IC-1	:	LM324 (Quad opamp)
S_1, S_2	:	ON/OFF switches

Microbuzzer, two 9V batteries, 14-pin DIL IC socket.

Assembly

INVISIBLE INTRUDER ALARM

Transmitter

Receiver

To Do and Notice

Now, switch S needs to be opened momentarily to reset the system and get it ready to detect the next intrusion.

What Happens?

In the transmitter, the current through the light emitting diode is being controlled by resistance R_1.

1   1

$_7$ does not

$_1$ goes to cut-off. Transistor

7

Try It Yourself

Project-37

Communicating with Laser Beam

How can a given electrical signal be transmitted from the one place to another riding over a laser beam ?

Introduction

We are all familiar with fiber optic cables that carry our telephone signals from one place to another. An optical fiber is a flexible, transparent fibre made of high quality extruded glass (silica) or plastic, slightly thicker than a human hair. It transmits light between the two ends of the fibre. The laser bream here is used as a carrier, which is modulated by the signal to be transmitted.

The basic principle of operation is the same whether we have the fibre optic link or the wireless link. The wireless laser communication links are very popular in space applications for providing inter-satellite communications.

Materials Required

R_1	:	4.7 KΩ, 1/4W carbon film resistor
R_2	:	2.2 KΩ, 1/4W carbon film resistor
R_3	:	1.2 KΩ, 1/4W carbon film resistor
R_4	:	33 KΩ, 2W carbon film resistor
R_5	:	470 KΩ, 1/4W carbon film resistor
R_6	:	820 KΩ, 1/4W carbon film resistor
R_7	:	68 KΩ, 1/4W carbon film resistor
R_8, R_9	:	1.5 KΩ, 1/4W carbon film resistor
P_1, P_2, and P_3	:	10 KΩ presets potentiometers
C_1, C_2, C_3, C_4, C_5, C_6, C_7, C_8, C_9 :		0.1 μF (ceramic disc) capacitors
C_2	:	0.1 μF (polyster/mica) capacitors
D_1, D_2	:	1N4001 diodes
LD-1	:	Laser pointer laser diode/any laser diode with about 1mW output power
PD-1	:	BPX-65 photodiode
IC-1	:	555 timer IC
IC-2, IC-3, IC-4	:	LF 356
B-1 to B-4	:	9V batteries
SW-1, SW-2	:	DPDT switch

Assembly

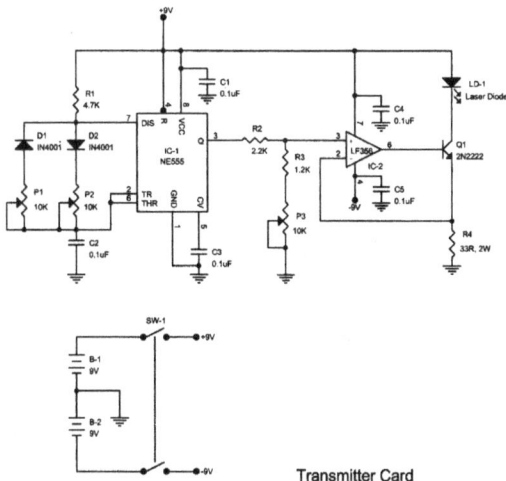

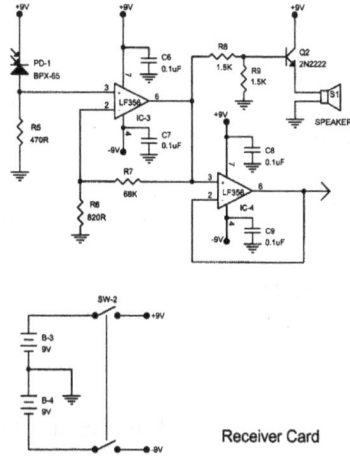

Transmitter Card

Receiver Card

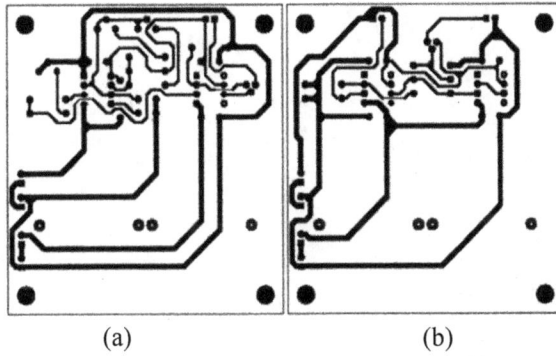

(a)   (b)

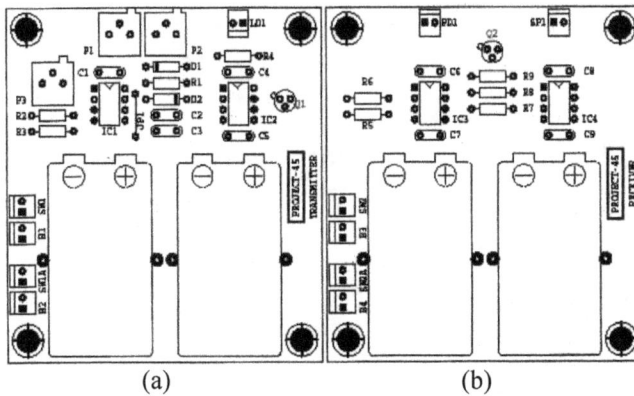

(a)   (b)

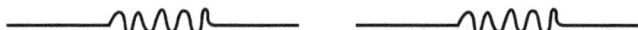

part.

To Do and Notice

desired signal is separated from the carrier by demodulation technique.

What Happens?

this pulse train can be set to the around 1kH[1]   [1]. The frequency of
                                                                    , R and P

                                    6   7   6

                                    . The unity gain buffer stage built[1])

for transmitter circuits.

Try It Yourself
Make laser-based perimeter protection system.

Project-38

Candles Go Electronic

Can you construct a simple project that can produce the effect of candle light (randomly flickering light) in a normal electric bulb ?

Introduction

This project is a simple circuit that can produce the effect of candle light in a normal electric bulb. A candle light as we all know resembles a randomly flickering light. We have to produce a randomly flickering light effect in an electric bulb.

Materials Required

R_1, R_2	:	10 KΩ, 1/4W carbon film resistor
R_3	:	180 Ω, 1/4W carbon film resistor
R_4	:	100 Ω, 1/4W carbon film resistor
P_1, P_2	:	100 KΩ, resets potentiometer
C_1	:	1 μF (tantalum) capacitor
C_2	:	0.01 μF (ceramic disc) capacitor
C_3, C_4, C_5, C_6, C_7, C_8, C_{10}, C_{11}	:	0.1 μF (ceramic disc) capacitor
C_9	:	1000 μF/25V DC electrolytic capacitors
D_1, D_2, D_3, D_4, D_5	:	1N4001 diodes
SCR-1	:	C106M or equivalent
VR-1	:	7805 voltage regulator
IC-1	:	555 timer
IC-2	:	76164
IC-3	:	7486
IC-4	:	7400
T-1	:	230V AC (primary) 7.5-0-7.5 V AC/250 mA secondary
L-1	:	Electric bulb, 40W to 100 W 230V AC
SW-1	:	ON/OFF switch

Assembly

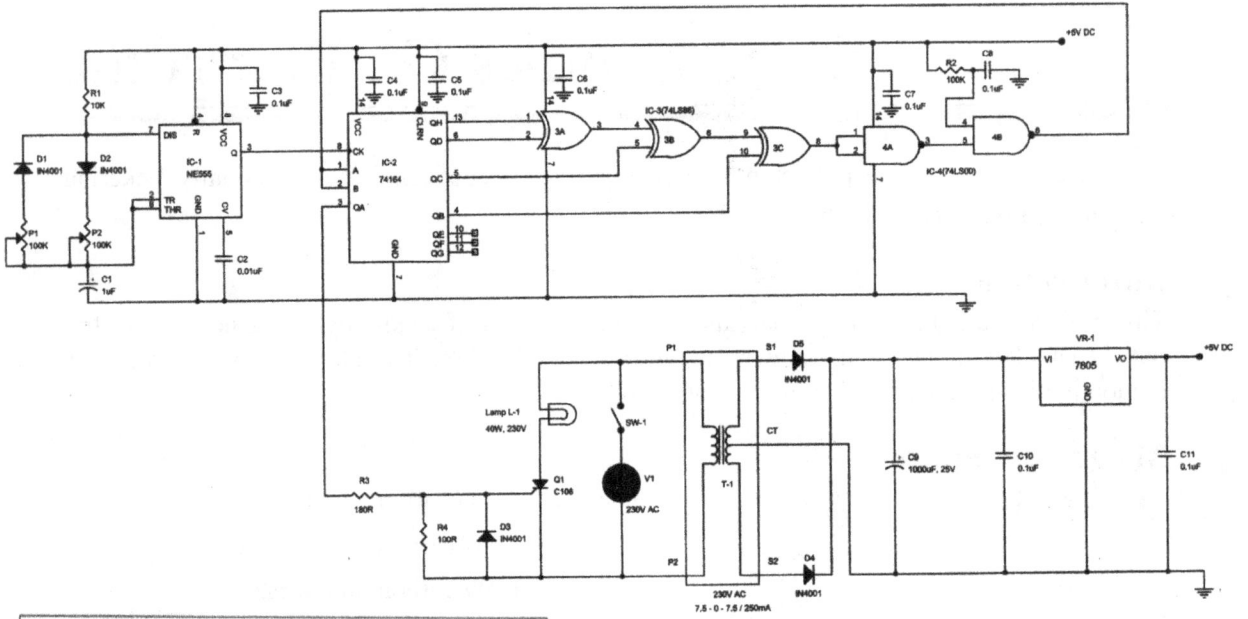

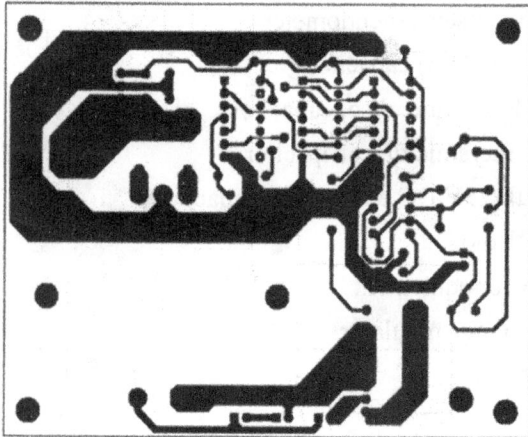

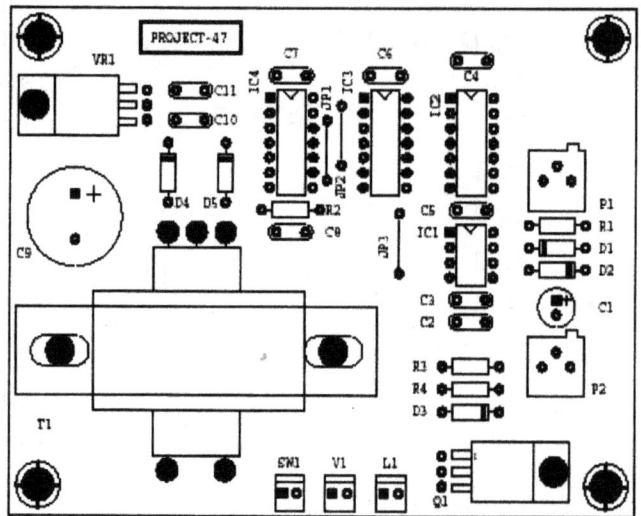

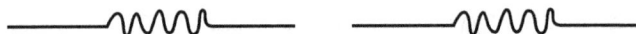

To Do and Notice

effect in the light output.

What Happens?

to get the best

light output.

Try It Yourself

Project-39

IC Timer Tester

How can you differentiate between good and bad ICs for a particular gadget ?

Introduction

The objective of this project is to build simple and easy to use test gadget that can be used to identify good and bad IC timers from a given lot. This project not only enables you to test the IC timer 555, it will also give you an opportunity to appreciate the tremendous function potential of various terminals of the truly versatile IC. It practically tests the functionality of each and every terminal of the timer with the result that a timer declared fit by this gadget is good in the true sense.

Materials Required

R_1, R_2	:	22 KΩ, 1/4W resistor (carbon film)
R_4, R_5, R_{11}	:	3.3 KΩ, 1/4W resistor (carbon film)
R_6	:	4.7 KΩ, 1/4W resistor (carbon film)
R_7	:	47 Ω, 1/4W resistor (carbon film)
R_8	:	150 KΩ, 1/4W resistor (carbon film)
R_9	:	47 KΩ, 1/4W resistor (carbon film)
R_{10}	:	1 KΩ, 1/4W resistor (carbon film)
R_3	:	10 KΩ, 1/4W resistor (carbon film)
P_1	:	1 KΩ, (preset) potentiometers
P_2, P_3	:	100 KΩ, (preset) potentiometers
C_1, C_3	:	0.01 µF (ceramic disc)
C_2, C_6, C_7	:	0.1 µF (ceramic disc)
C_4	:	10 µF, 25V (electrolytic)
C_5	:	100 µF, 25V (electrolytic)
D_1, D_2, D_3, D_4	:	1N4001 or equivalent diodes
LED-3, LED-1, LED-2	:	Preferably of different colours
IC-1	:	CD4011B
SW-1	:	Microswitch
SW-2, SW-3	:	DPDT switch

Assembly

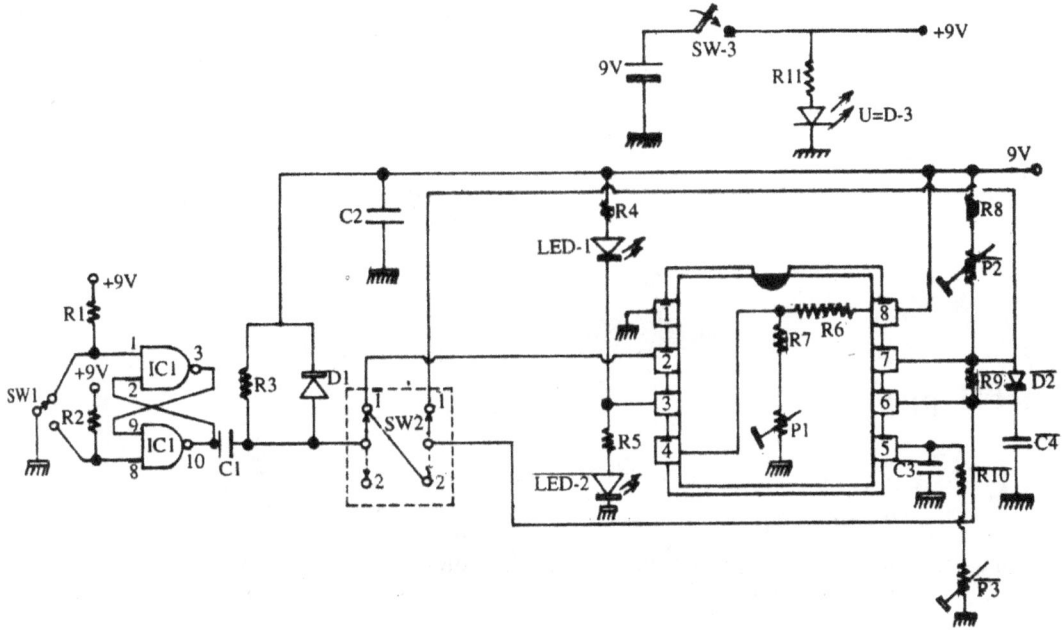

the microswitch. Resistor R$_1$ to ground trigger pulse. This differentiator circuit pulse is less than the expected monoshot output pulse. Remember the trigger pulse width which is less than the expected monoshot output pulse which is an essential requirement of this mode of operaton.

The output is a pulse train with the high time determined by the total resistance offered by the series combination of R_8, R , P   whereas the low time is determined by resistor R   .

in the normal circumstances. More

pulse train in the astable mode and the pulse width at the output in monoshot mode by applying

To Do and Notice

be checked.

What Happens?

train in astable mode and the pulse width at output in monoshot mode.

1.1 × (total charging resistance) × (charging capacitance)

Try It Yourself

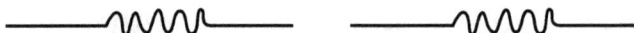

Project-40

OP-AMP Tester

How can you test an opamp type 741 ?

Introduction

This test gadget can be used to test for general purpose opamp type 741 and other such opamps that are pin to pin compatible to 741. These include the opamp type numbers such as LM 709, LM 201, MC 1439, LM 748, OP-02, LM 318, LM 356 etc. With this gadget, we do not have to randomly replace the opamps in case the circuit we have built with so much enthusiam happens to use a few opamps and is not working. We will be able to test the opamp before declaring it fit for use.

Materials Required

R_1	:	470Ω, 1/2W carbon film resistor
R_2	:	1KΩ, 1/4W carbon film resistor
R_3	:	1Ω, 1/4W carbon film resistor
R_4	:	1KΩ, 1/4W carbon film resistor
R_5	:	2.2KΩ, 1/4W carbon film resistor
R_6	:	2.2KΩ, 1/4W carbon film resistor
C_1, C_2	:	100 µF/35 V electrolytic capacitors
C_3, C_4, C_5, C_6	:	0.1 µF ceramic capacitors
IC-1	:	Three terminal voltage regulator type 7812
IC-2	:	Three terminal voltage regulator type 7912
D_1 to D_4	:	IN4001 or equivalent diodes
LED-1 and LED-2	:	The two LEDs should preferably be of different colours
VZ-1	:	3V, 400 mW zener diode
T-1	:	Mains transformer (15-0-15, 250 mA)
SW-1	:	ON/OFF toggle switch
SW-2	:	DPDT switch
Fuse	:	500 mA
IC socket	:	8-pin DIL type

Assembly

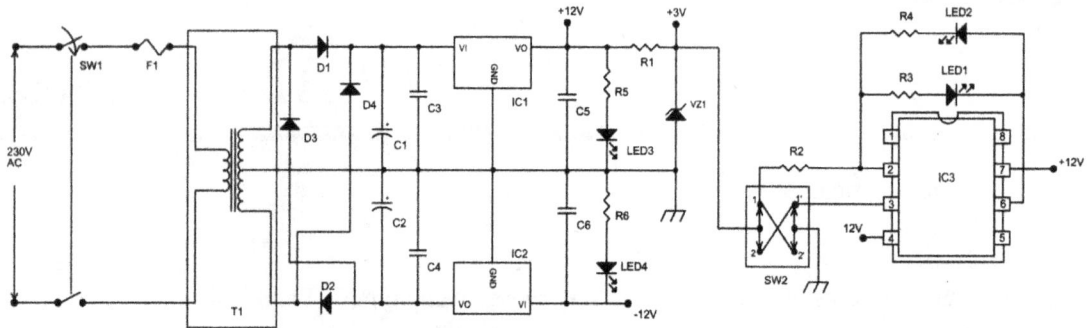

1

1•

does the

1

To Do and Notice

7. Switch off the power supply.

What Happens?

Try It Yourself

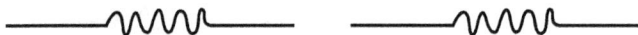

Project-41

Transistor Tester

Can you make a tester circuit to test the bipolar junction transistor both PNP and NPN ?

Introduction
This project is a simple circuit that can be used to test BJT transistors. In addition to testing the transistors for any defect. This easy to build test gadget can also be used to measure the (hFE) at a given healthy transistor.

Two different sets of three terminals each have been brought out on the front panel, one for testing NPN and the other for testing PNP transistors. Also (hFE) measurement can be carried out for a wide range of transistors starting with low (hFE) power transistors to high (hFE) small signal transistors. An (hFE) as high as 1000 can be measured with this gadget.

Materials Required

R_1	:	1KΩ, 1/4 W carbon film resistor
R_2, R_3	:	470KΩ, 1/4 W carbon film resistor
P_1, P_2	:	100KΩ, preset potentiometers
VZ-1, VZ-2	:	5.6V, 400 mW zener diodes
Q_1	:	2N2907 PNP transistor
Q_2	:	2N2222 NPN
S_1	:	ON/OFF switch
Battery	:	9V battery

Assembly
The heart of the circuit has two constant current sources built around transistors (Q_1) and (Q_2). Q_1 is a PNP transistor.

To Do and Notice
If the transistor under test is NPN, then provide current by Q_1 circuit. If the transistor under test in PNP provide current by Q_2-circuit. The constant current is fed to the base of transistor under test. This current is multiplied by the (h_{FE}) of the transistor under test and flows in the collector which is

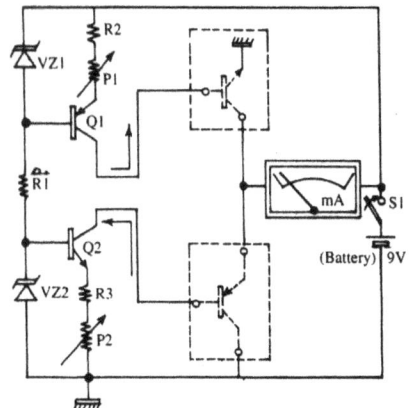

Fig. 41.1 : Circuit Diagram

indicated by the meter.

This is true as the transistor under test is always operaing in the non-saturated mode.

What Happens?

1

———

P_1

———

)

directly calibrated to read the (h   ) of the transistor. The (h
for the functioning of the transistor.

Try It Yourself

transistors.

Project-42

Variable Power Supply and Charger

Can you make a mobile phone battery charger with emergency light ?

Introduction

This is very easy circuit of 'variable power supply and charger'. It is not very much useful in the time of power cut but can be used as main power supply. At your workbench you can use this circuit to check or test your electronic projects. Mobile phone batteries can be changed with the help of these circuits.

This circuit can also work as an emergency light.

Materials Required

IC	:	LM 317
R_1	:	220Ω, 1% carbon film resistor
R_2 to R_{12}	:	220Ω, 1% carbon film resistor
R_{13}	:	470Ω, 1% carbon film resistor
VR-1	:	100KΩ variable resistor
C_1	:	100 µF capacitor
C_2	:	0.1 µF capacitor
D_1 to D_4	:	IN4005 diodes
S_1 to S_5	:	ON/OFF switches
LED_1 to LED_{12} :		LEDs
Zener diode	:	3.3V
Transforer and battery		

Assembly

As per your need you can take the output from the circuit by just flipping different number of switches (from S_3, S_4 and S_5) in the circuit. S_1 and S_2 are the two switches that are given in the circuit so that you can power your circuit either directly with the AC supply or else you can take help of any battery.

D1-D4=1N4007
R2-R12=220E
S1-S5=ON/OFF
SWITCH

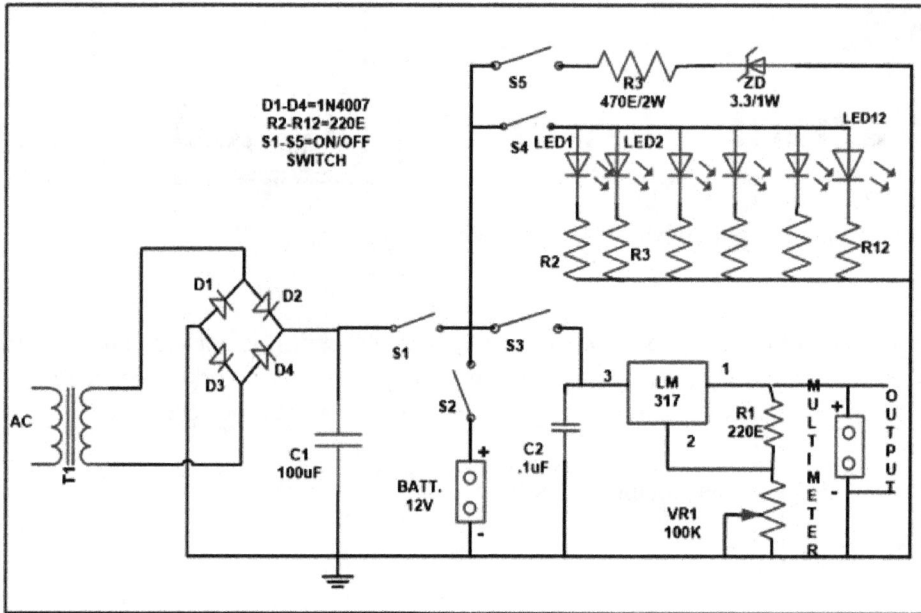

To Do and Notice

to switch S if you want to use the emergency light.

of power failure.

What Happens?

[1] while if you want

Try It Yourself

Try to make an automatic street light controller circuit.

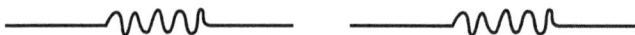

Project-43

USB Mobile Charger Circuit

Can you make a USB mobile charger ?

Introduction

We can use this circuit to charge a mobile while we are going on a journey. So, we may treat it as mobile travel charger circuit.

The USB part of the mobile is used for charging as the USB port is very helpful voltage source that can charge the mobile. Nowadays, there are two to four USB ports on the laptops that are available in the market. USB actually refers to Universal Serial Bus. It is the newest method which is used to get information in and out from your computer. We are concerned with the fact that ± 5V of power is being provided by the USB port to external devices and can avail at the pin-1 while on pin-4, it is 0 V, till 100 mA of current we can get from the USB port which is more than sufficient that we require for this small application.

Materials Required

R_1	:	470 Ω, 1% resistor
C_1	:	100 µF/25V capacitor
T_1	:	BC 547 transistor
Zener diode	:	4.7V/0.5 W
Diode	:	1N4007
USB	:	Cables

Assembly

The figure 43.2 shows the circuit diagram. Resistors are used to control the circuit current. Capacitors are mainly used to store charges, electrolytic capacitor is an example of polarized while ceramic and paper capacitors are non-polarised. Zener diode starts working at breakdown point. It works always in reverse bias state.

The circuit explained above works on 4.7 regulated voltage. The voltage at output is harmonised by the transistor named T-1.

Fig. 43.1 : USB Port

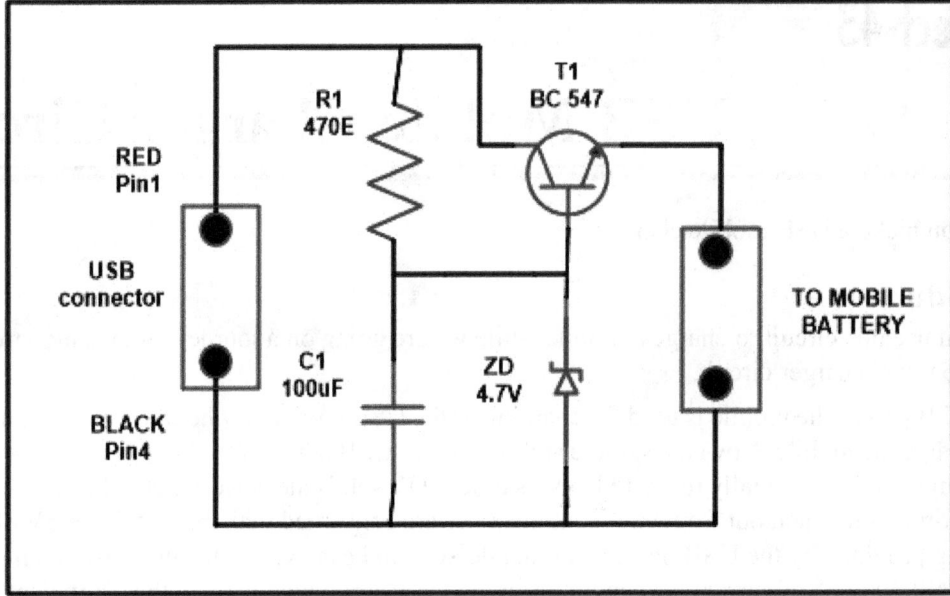

To Do and Notice

What Happens?

Try It Yourself

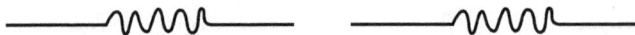

Project-44

Wireless Switch

Can you design a circuit by which any electrical appliance make ON/OFF without physical contact with the appliance ?

Introduction

Generally, appliances that we use in our home are being controlled with the help of devices like switches, sensors. However sometimes it is dangerous to have physical contact with these switches. So, to overcome these dangers, we have explained a circuit that needs no physical contact with the appliance. In this circuit all you need to pass is your hand above LDR. As you first pass your hand over LDR, the device connected with it starts and remains in that state till you again pass your hand above LDR.

Materials Required

IC_2	:	CD4027, J-K flip flop
IC_1	:	LM741 opamp
R_1	:	1KΩ, carbon film resistor
R_2	:	82Ω, carbon film resistor
R_3	:	10Ω, carbon film resistor
VR-1	:	5MΩ, variable resistor
Relay	:	Relay switch
D_1	:	1N4001 diode
LDR	:	LDR
T_1	:	BC547 transistor
C_1	:	0.1 µF
Battery	:	6V battery

Assembly

This circuit mainly depends on two ICs. First one is LM741 which is an opamp amplifier. This IC is used to increase the voltage level at the output hundred of a thousand times as comapred with the input. Another IC is most generally used J-K flip flop. This IC works in toggle mode and is based on the J-K flip flop master slave concept.

It is used to alter the state when the signal is given to the any one of the input terminals and can get more than single output. These are four input pins J-K flip flop named J and K along with set and

— for output.

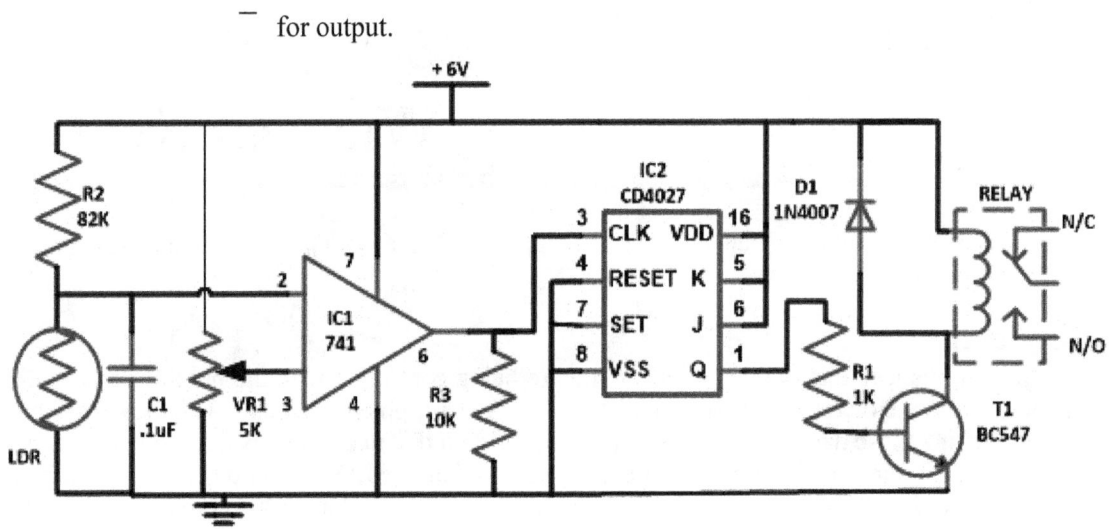

To Do and Notice

What Happens?

[1]

[1]. The logic

[1]

[1]

Try It Yourself

Project-45

Fire Alarm

Making a fire alarm system

Introduction

Fire Alarm alerts us when there is a fire accident at home by ringing a siren sound. We are aware that there are many integrated circuits which can be used to generate the siren effect but we prefer to use basic elecronic components like resistors, capacitors and transistors to generate it so that you will clearly understand the internal working of it and it will be much useful for you as you will gain more knowledge by alayzing it instead of simply pre-designed integrated circuits.

Materials Required

TH-1	:	Thermistor
RV-2	:	Potentiometer arrangement
R_2	:	4.7 KΩ
R_1	:	120 Ω
R_5	:	10 Ω
R_3	:	4.7 KΩ
R_4	:	4.7 KΩ
R_6	:	56 KΩ
R_7	:	470 KΩ
R_8	:	22 KΩ
R_9	:	39 KΩ
R_{10}	:	47 KΩ
R_{12}	:	470 KΩ
Q_1, Q_2	:	BC547 transistor
Q_3	:	BC107 transistor
Q_4	:	2N2907 transistor
Q_5	:	2N2222 transistor
C_1	:	22μF
C_2	:	0.47

Assembly

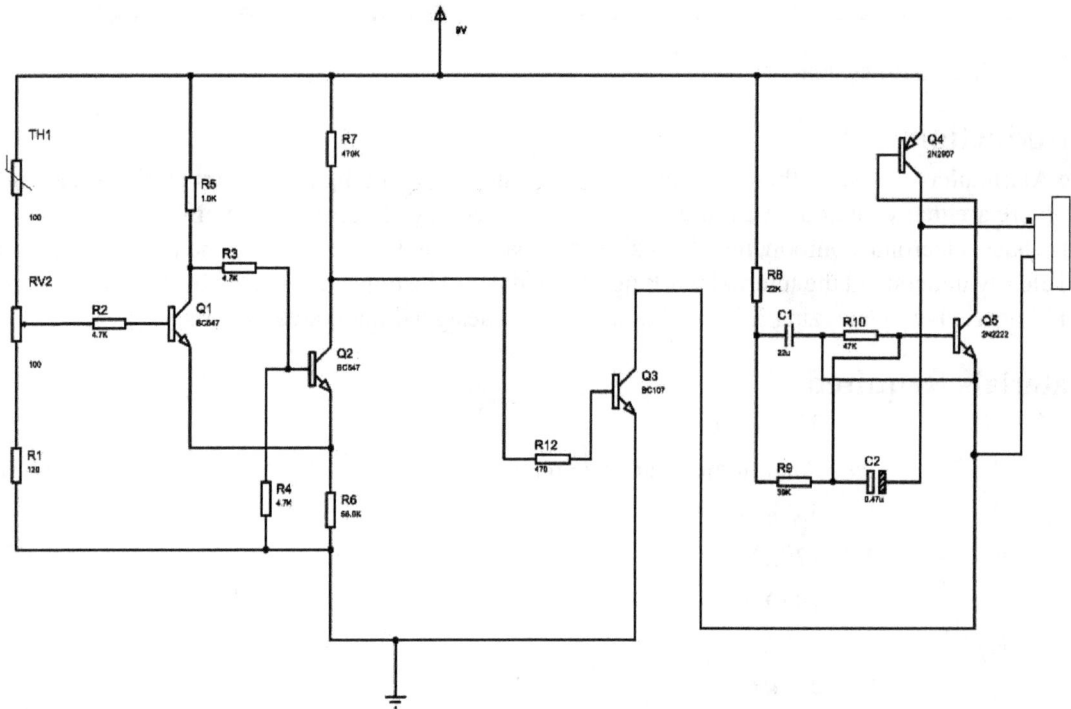

To Do and Notice

What Happens?

too in common emitter mode and as the input is low when the temperature threshold is reached, the transistor will now act as a switch for the siren circuit. This transistor can bear power quite larger

The capacitors used in the circuit are the main components in producing siren effect.

Try It Yourself
Try to design a Panic alarm circuit.

Project-46

Metal Detector

How to design a circuit by which any explosive metals or illegal things like guns can be detected ?

Introduction

Metal dectctor is very common device for checking the person in shopping malls, hotels, cinema halls, to ensure that a person is not carrying any illegal explosive metals like guns, bombs etc. Metal detectors can be created easily and its circuit is not complex.

Materials Required

IC	:	TDA 0161
R_5, R_4, R_1	:	1 KΩ
R_2	:	330 Ω
R_3	:	120 Ω
RV-1	:	Variable resistor 10 kΩ
C_1, C_2	:	0.047 μF
BT-1	:	4V battery
SW-1	:	SPST key
Q_1	:	2N3904 transistor
Z-1	:	3V buzzer
D_1	:	LED
LC-circuit	:	680 PF

Assembly

As the fig. 46.1 shows the circuit diagram and fig 46.2 shows the block diagram. IC TDA0161 is a proximity sensor which can detect the objects without any physical interference. It works same as infrared sensor, proximity also releases a signal. It will not give output unless and until there is no change in the reflected back signal. If there is charge in signal, it will detect and give the output accordingly. LC circuit is a resonating circuit which will resonate when exact same frequency material comes near. The LC circuit consists of inductor and capacitor connected in parallel. When the capacitor is fully charged, the charge of the capacitor will be given to the inductor, here inductor will improve its magnetic field.

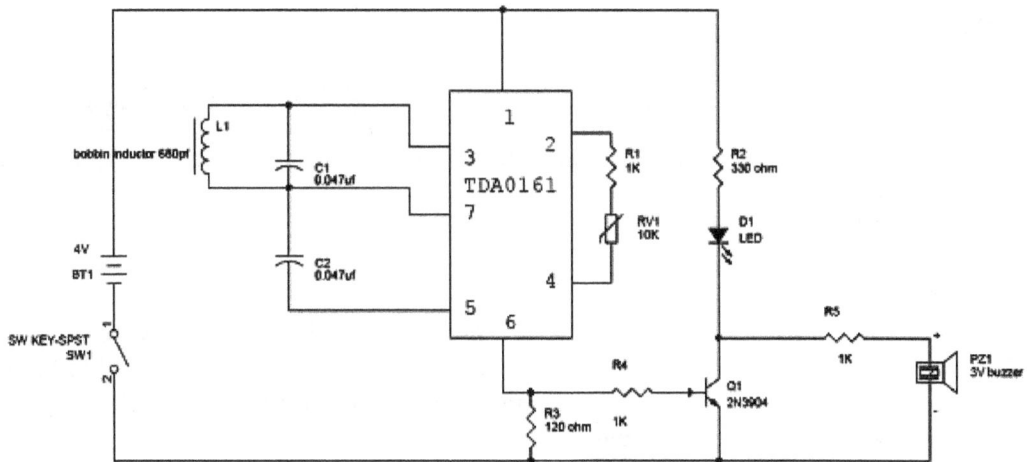

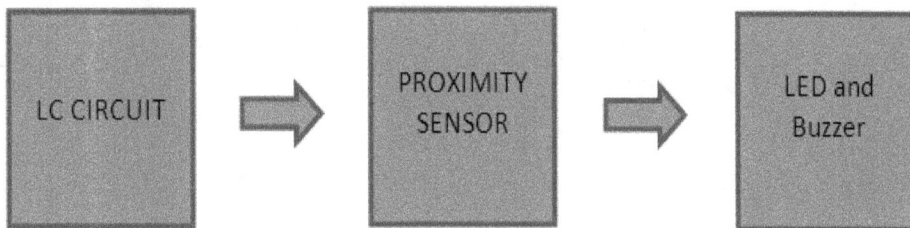

To Do and Notice

|

What Happens?

| has got any resonating frequency from any metal which is

proximity detector, which will detect the charge in the signal and react accordinly. The output of

when coil is near the metal.

Try It Yourself

Project-47

Mobile Jammer

How to make a restricted area in which cell phones can not work ?

Introduction

By the use of a mobile Jammer circuit, we can make or cell phone restricted area. This circuit will work in the range of 100 meters i.e. it can block the signals of cells phones within 100 meters radius. This circuit can be used in TV transmission and also for remote controlled toys or play things. The usage of this type of circuits is banned in most of the countries.

Materials Required

R_1	:	39 kΩ
R_2	:	100 kΩ
C_1	:	15 pF
C_2	:	4.7 pF
C_3	:	4.7 pF
C_4	:	102 pF
C_5	:	1 µF
C_6	:	2.2 pF
C_7	:	10 pF
L_1	:	22 nH
Q_1	:	BF 494 transistor
Battery	:	3V battery
Antenna	:	Yagi uda antenna

Mobile Jammer

Assembly

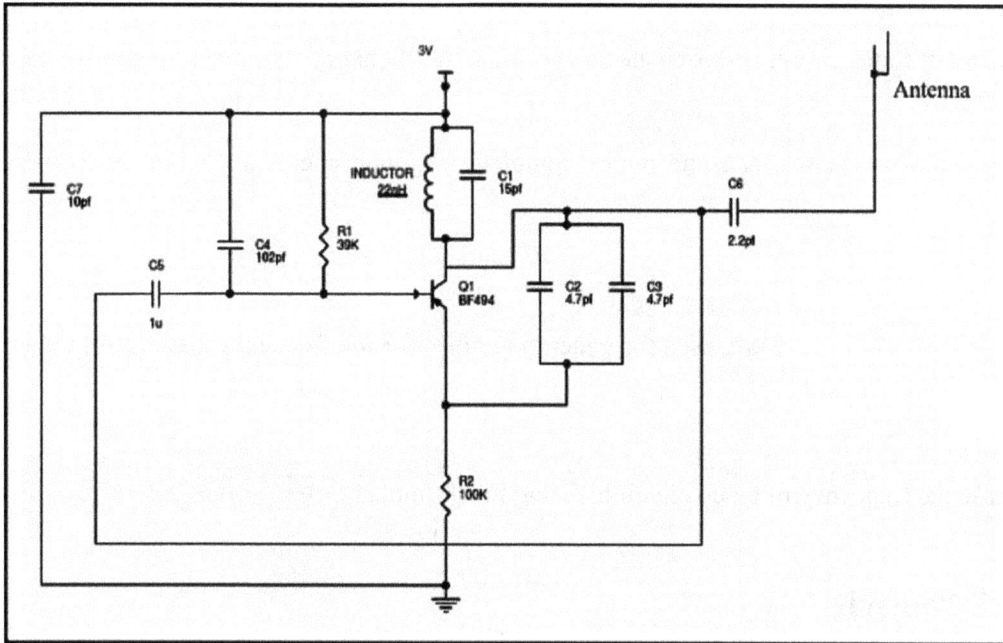

(ii) Voltage controlled oscillator
(iii) Tuning circuit

₁   and resistor R₁

₁   ₁ constitute of tuned circuit.

To Do and Notice

can block the cellphone signal from reaching the cell phones.

What Happens?

capacitor of tuned circuit will oscillate at its resonating frequency. The main function of capacitor

the capacitor decreases, at some points complete magnetic energy is stored by inductor and

to the

₆ gain signal.

are used for generating noise for the frequency generated by the tuned

increase the frequency of tuned circuit by using this formula.

Try It Yourself

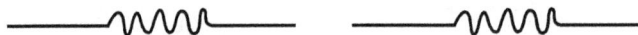

Project-48

FM Jammer

How do you make a restricted area in which any FM radio can not work ?

Introduction

In old days when we used analog signal for communication, the Jamming circuit was easily used for producing high frequency noise signals, but today the trend is completely changed and the use of digital devices has taken place of analog devices. High frequency signals are not capable to block those signals from reaching the devices, so we need very high frequency signals to block actual signals from reaching the devices. The Jammer circuit produces high frequency noise signal which will confuse the receiver of a particular system from receiving the signal. User of the system feels that the circuit is not working properly.

Materials Required

R_1	:	15 KΩ resistor
R_2	:	3.9 KΩ resistor
R_3	:	220 KΩ resistor
C_1	:	Variable capacitor of 6 to 35pF
C_2	:	6 pF
C_3	:	0.01 pF
C_4	:	10 pF
Q_1	:	2N2222 transistor
L_1	:	Inductor
BT-1	:	9V battery
Antenna	:	Yagiuda antenna

Assembly

Capacitor C_1 and L_1 will constitute tank circuit. Transistor Q_1 will start the operaton of tank circuit. Resistors R_1 and R_2 will act as the biasing circuit and R_3 is used for limiting the emitter current in the circuit.

To Do and Notice

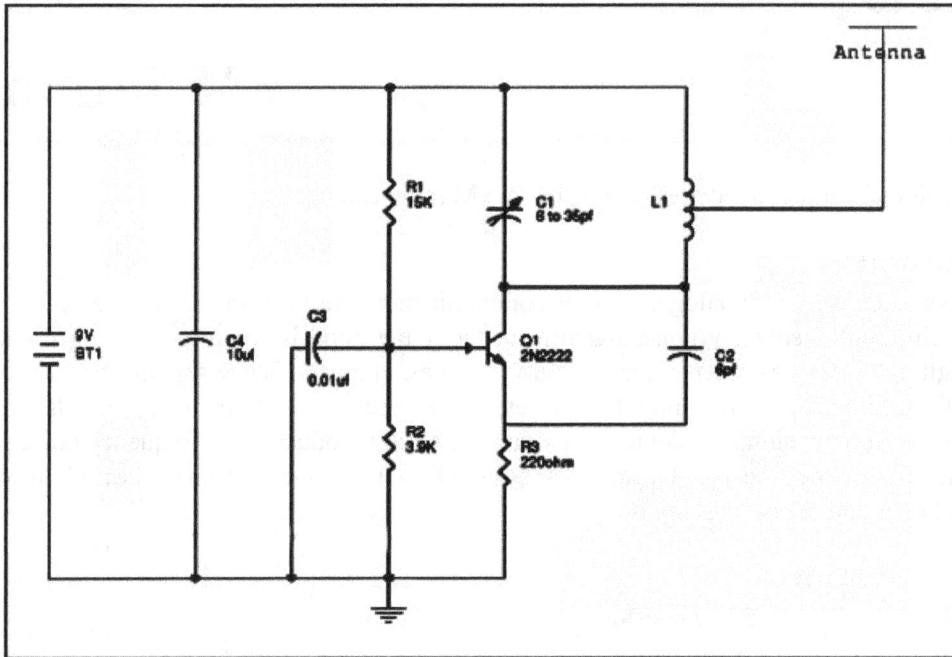

What Happens?

[1]

[1]

formula

Try It Yourself
Try to make a motion detector circuit.

Project-49

Railway Safety System

Can you make an automatic railway gate controller at the unmanned railway crossing with high speed alerting alarm system ?

Introduction

The main aim of this project is to operate and control unmanned railway crossing properly in order to avoid accidents. In a country like ours where there are many unmanned railway crossings, accidents are increasing day by day. These train accidents occur due to absence of human power in the railway. In order to overcome the accidents, the Railway Safety System is designed and installed at unmanned Railway crossing.

It has detectors at at a far distance on the railway track which helps us to know the arival and departure of a train. These detectors are attached to micro controller which activates motors and opens/closes the railway gate.

Another feature of this circuit is that it has an intelligent alerting system which detects the speed of the train that is arriving. If the speed is found to be higher than the normal speed, then micro controller automatically activates the alarm present at the gate.

Materials Required

10 kΩ variable resistors	:	4
100 kΩ variable resistors	:	4
2.2 kΩ variable resistors	:	1
101 Ω variable resistors	:	1
100 Ω variable resistors	:	1
10 kΩ variable resistors	:	1
BC 547 transistor	:	1
AC motor, 240 VAC	:	1
6 V DC Relay switch	:	1
Photodiodes	:	4
IR LEDs	:	4
ATMED A16L micro controller	:	1
470 µF/25V capacitors	:	2
10 µF/16V capacitors	:	2

Assembly

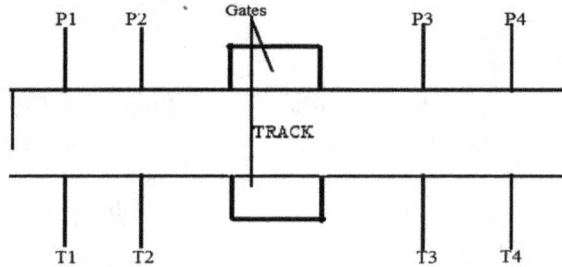

Sensors
T- IR LED (Transmitter)
P-Photodiode(Receiver)

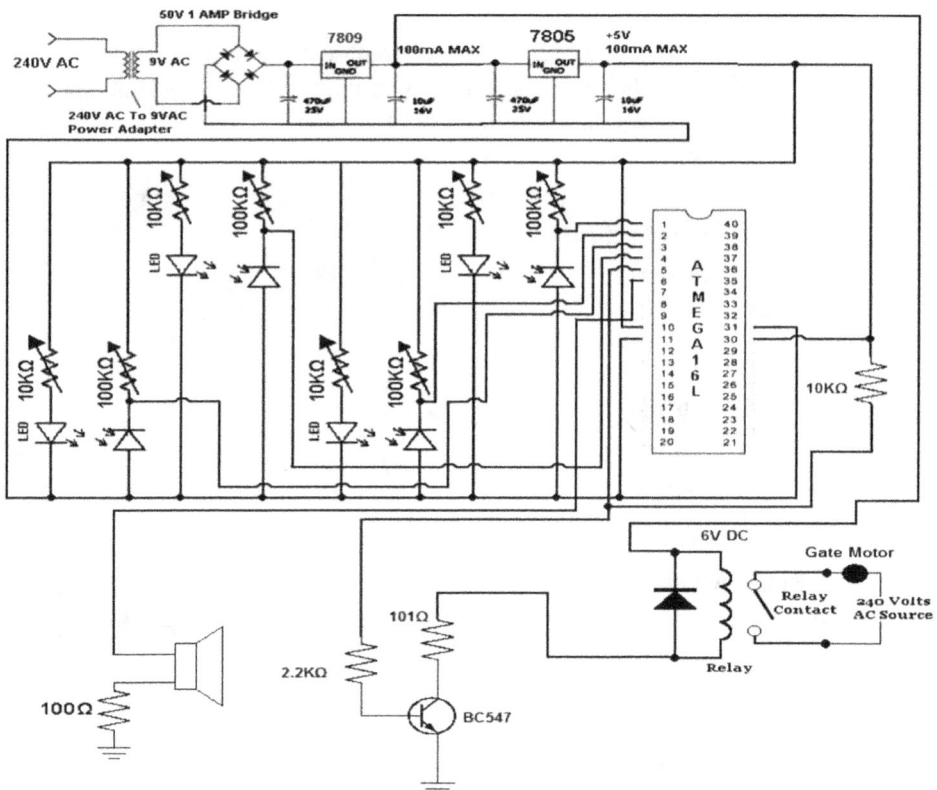

To Do and Notice

opened.

What Happens?

track at about 1 meter apart, with transmitter and photodiode of the transmistter and the photodiode of each pair on the opposite sides of the track. The system displays the time taken by the train to

This circuit has been designed considering the maximum perimissible speed of trains as per the

desired output appropriately.

Try It Yourself

Try to comprehend how a sensor detects the last carriage.

Project-50

FM Bugger

Can you listen the conversation of another person from a long distance using the normal FM radio set ?

Introduction

Bugger is a device which gives information of one person to the other in remote location. Normally bugg is used to find out the status of a person i.e. where he is going, what he is talking etc. Here is a small circuit with which you can listen to another person's conversation from long distance using FM radio set. The FM bugger circuit is kept in room where you want to listen the conversation. The range of transmission area is around 100 meters.

Materials Required

R_1	:	22 kΩ, resistor (carbon film)
R_2	:	47 kΩ, resistor (carbon film)
R_3	:	330 Ω, resistor (carbon film)
L_1	:	Inductor coil
C_4, C_2, C_1	:	1 nF
C_3	:	4.7 pF
C_5	:	22 nF
SW-2	:	ON/OFF toggle switch
Battery	:	3V battery
C_6	:	50 pF variable capacitor
MIC	:	Microphone
Q_1	:	2N2222 transistor
Antenna	:	Dipole antenna

Assembly

Transmitter Section:

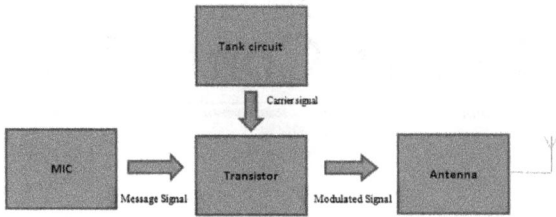

Receiver Section:

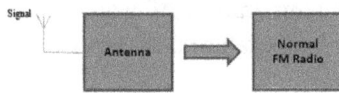

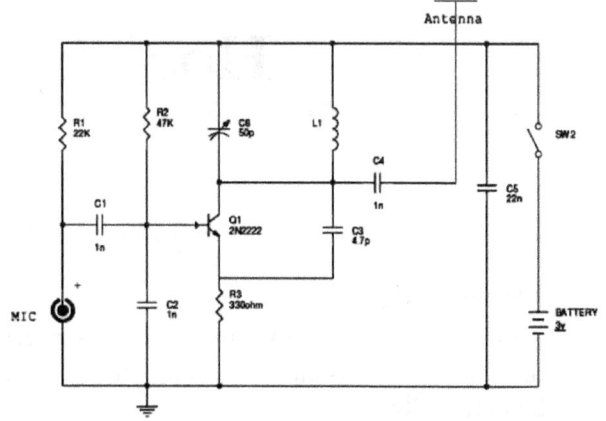

1

The circuit uses analog modulation in which the carrier signal is applied continuously to the

and is modulated to the carrier signal and transmitted.

To Do and Notice

What Happens?

in accordance to instantaneous amplitude of modulating signal.

1

6   1

the noise in the transmitted signal.

6

Try It Yourself

Project-51

Hot Water Geyser Controller

Can you make a circuit which turns a geyser OFF/ON when water gets hot and ready for bathing?

Introduction

This circuit is made to turn ON/OFF the geysers when the temperature of water reaches at desired value. This circuit has an inherent ability to detect the temperature and turn ON or OFF automatically.

This circuit has thermistor which is a temperature sensing element and helps the circuit to determine the temperature of the water.

Materials Required

R_1	:	120 Ω
RV-2	:	Variable resistor 100 kW
R_2	:	4.7 kΩ
R_4, R_2	:	4.7 kΩ
R_5	:	1.0 kΩ
R_6	:	56 kΩ
R_7	:	470 kΩ
R_8	:	100 kΩ
R_{12}	:	470 kΩ
Q_1, Q_2	:	BC 547 NPN transistor
Q_3	:	BC 107
RL-1	:	Relay switch
Battery	:	9V battery
TH-1	:	Thermistor

Assembly

The thermistor is kept in contact with water while assembling the circuit on field. Transistor BC 547 is a general purpose NPN transistor. Here, transistors are used as a switch. It drives the relay switch.

TH1

R5
1.0K

R7
470K

RV2

R3
4.7K

R2
4.7K

Q1
BC547

100

Q2
BC547

RL1
OMIH-SH-105L

To Geyser Switch

R8
100R

Q3
BC107

R1
120

R4
4.7K

R6
56.0K

R12
470

9V

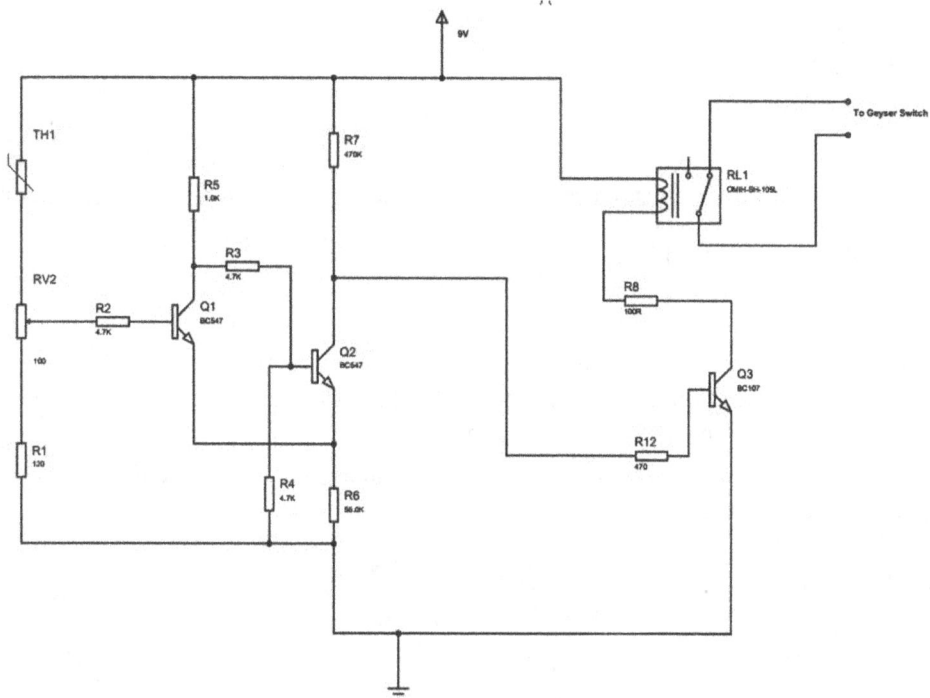

To Do and Notice

What Happens?

and the geyser will remain

relay turns   which further breaks the connection and so the geyser turns   .

Try It Yourself

Project-52

Battery Charger

Can you make an Lead-Acid battery charger ?

Introduction

To charge a battery from AC, we need a step-down transformer, rectifier, filtering circuit, regulator to maintain the constant voltage that we can give to the battery to charge it. Suppose you have only DC voltage and want to charge the lead acid battery.

The main advantages of lead acid battery is its working efficiency and its durability. It can deliver high current at very low cost.

Materials Required

U_1	:	LM317/CYL voltage regulator
R_1	:	100 Ω
R_2	:	0.5 Ω
R_3	:	120 Ω
R_4	:	470 Ω
R_5	:	1 kΩ
D_1	:	LED
C_1	:	100 μF
C_2	:	0.22 μF
C_3	:	0.22 μF
D_2	:	Diode
Q_1	:	BC 547

Assembly

As seen in fig. 52.1, DC voltage is given to the DC voltage regulator LM317. The regulated DC output voltage is given to battery. There is also a trickle charge mode circuitry which will help to reduce the current when battery is fully charged.

Fig. 52.1 : Block Diagram of Charger for Lead Acid Battery

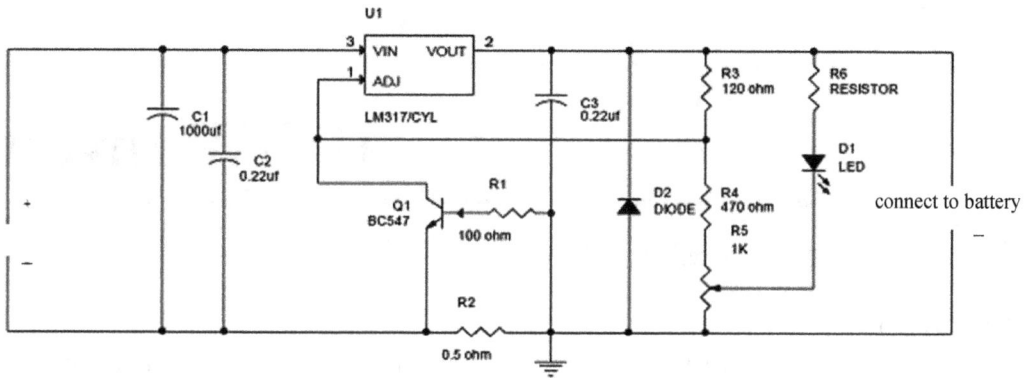

To Do and Notice

regulator will reduce the charge current.

What Happens?

R_1, R , R [1]. Resistor

           [1] resistor R_1 and R and potentiometer R .
Potentiometer is used to set the charging current. Resistor R

[1].

Try It Yourself

Project-53

100 W Inverter

Can you design a 100 W inverter circuit ?

Introduction

Inverter is a small circuit that will convert the direct current (DC) to alternating current (AC). The power of battery is converted into main voltages or AC power. This power can be used for electronic appliances like television, mobile phones, computer etc.

The main fuction of the inverter is to convert DC to AC and the step-up transformer is used to create main voltage from resulting AC.

Materials Required

R_1	:	330 Ω, carbon film resistor
R_5	:	1 kΩ, carbon film resistor
R_{10}	:	250 Ω, carbon film resistor
R_8	:	1 kΩ, carbon film resistor
R_9	:	1 kΩ, carbon film resistor
BT-1	:	12V Battery
C_3	:	220 μF/25V
C_1	:	0.01 F
C_2	:	0.1 μF
D_5, D_6	:	Zener diode
D_2	:	LED
D_3	:	Diode (1 N4007)
SW-2	:	ON/OFF switch (SPST)
U_1	:	CD 4047 multivibrator
Q_1, Q_2	:	IRF 540 N-channel enhanced mode MOSFET
T_2	:	Step-up transformer

Assembly

To Do and Notice

What Happens?

is V

duty cycle. The output frequency is connected to the   through resistor which will help

Try It Yourself

Project-54

Continuity Tester with Melody

How to design a circuit which can test broken wire i.e. continuity and undesired shorting of wires ?

Introduction

Many a times when we are connecting the components on the PCB or bread board, there is a possibility of compnents to get attached whether due to defects in the PCB or bread board or due to mistakes committed during assembling of the circuit. These defects are minor and superficial and are in the form of breaks in the wires. The small, inexpensive circuit of a continuity tester may be used to detect such defects.

Materials Required

R_1, R_2	:	1 kΩ carbon film resistor
U_1	:	IC 555
U_2	:	IC UM66 (music generating IC)
LS-1	:	Loud speaker 8Ω
Q_1	:	BC 107 transistor
C_1	:	0.01 8 µF capacitor
D_1	:	1N4728A capacitor
Battery	:	9V battery

Assembly

Here 555-IC works in astable mode of operation. 555-IC is designed to generate 2 KHZ output frequency. Output of 555-IC is connected to 8 ohms speaker. Reset pin is connected to V_{cc}. Here 4th pin is connected to ground through 1.0 KΩ resistor, so normally it is OFF. When probes connect to the testing point, 4th pin will get positive V_{cc} and speaker starts sound. Potentiometer helps to set the resistance to determine the point when the buzzer should turn ON.

To Do and Notice

If you want to check if the wire is connected from one end to another, then use the probes of continuity tester and put to the ends of the wires to be tested. If the wires are connected, then the circuit makes a sound indicating that the wire is continuous without any breakage in the middle.

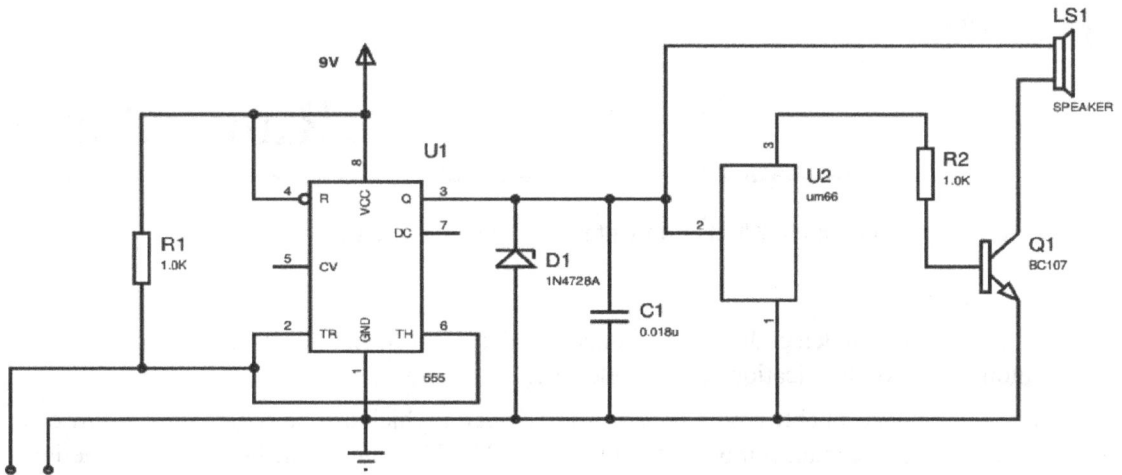

What Happens?

The output is high only when the circuit detects that the probes are started. If not, the output is

The loud speaker is a mini 8
this circuit.

Try It Yourself

Project-55

Rain Alarm

How to make a project which detects rain water and makes us alert ?

Introduction

Rain water detector will detect the rain and make an alert. This detector is used in irrigation fields, home automation, communication, automobiles etc.

Here is the simple and reliable circuit of rain water detector which can be constructed at low cost. Rain water sensor is the main component in the circuit. NE 555 timer is used to control the actions.

Materials Required

R_1, R_4	:	470 kΩ carbon film resistor
R_2	:	270 kΩ carbon film resistor
R_3	:	10 kΩ carbon film resistor
R_5, R_6	:	3.3 kΩ carbon film resistor
R_7	:	68 kΩ carbon film resistor
C_2, C_4	:	100 μF capacitor
C_1	:	220 μF/12V
C_3	:	0.01 μF
Q_1, Q_2, Q_3	:	BC148 transistor (NPN)
D_1	:	LED
D_2	:	1N4148 transistor
Battery	:	9V battery
U1	:	NE 555 timer IC
LS-1	:	Loud speaker 8Ω

Assembly

As seen in fig. 55.3 the rain sensor is the main component in the circuit. For the rain sensor, take the piece at bakelite or mica board and aluminum wire. Now, Aluminum wire should be posted on the flat board as shown in figure at rain sensor. This sensor is connected to the circuit i.e. between probes A and B.

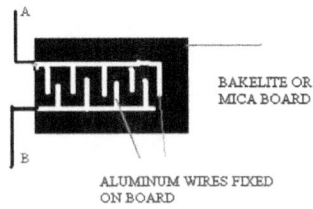

BAKELITE OR
MICA BOARD

ALUMINUM WIRES FIXED
ON BOARD

RAIN WATER
SENSOR

NE555
TIMER

BUZZER

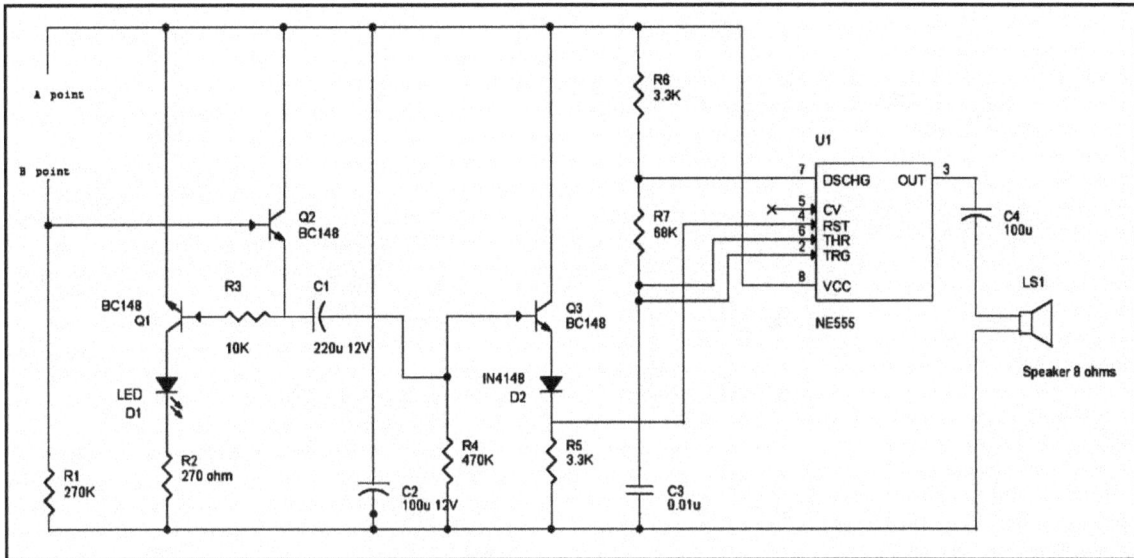

To Do and Notice

between the wires in the sensor. Its raining the water droplets will fall on the rain sensor which will
decrease the resistance between the wires and wires on the sensor board will conduct and trigger

What Happens?

will be turned

$_1$ will get charged by the resistor R

reaches the saturation mode.

will

$_1$

stop making sound.

Try It Yourself

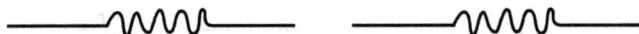

Project-56

Boolean Algebra Calculator

How to design a circuit which calculates the Boolean algebra problems ?

Introduction

The Boolean algebra calculator gives a basic logic for the operations on binary numbers 0 and 1. This boolean calculator is well known as boolean algebra simplifier or boolean algebra solver. In our circuit, we use Boolean algebra simplification methods like the Quine McCluskey algorithm to simplify the Boolean expression and display the output.

It works as a portable calculator to simplify the Boolean expression on the fly.

Materials Required

R_1, R_2, R_3, R_4, R_5, R_6, R_7, R_8, R_9	:	3.3 kΩ carbon film resistor
R_{11}	:	1 kΩ
R_{10}	:	10 kΩ variable resistor
S_1, S_2, S_3, S_4, S_5, S_6, S_7, S_8	:	ON/OFF switch
LED_1, LED_2, LED_3, LED_4, LED_5, LED_6, LED_7:		LEDs
IC_1	:	7805 voltage regularor
Battery	:	9V battery (DC)
IC_2	:	ATMEGA 16 L microcontroller

Assembly

Fig. 56.2 shows the block diagram of Boolean expression minimization calculator. Block diagram shows the 4 parts of our project which are

(i) **Power supply :** In our project a supply mains that is 5V DC is given to micro controller, LEDs, keypad and display.

(ii) **Micro controller :** ATMEGA 16 L is used for the automation purpose and acts as brain of the project. It controls the output (display) according to the input given to it.

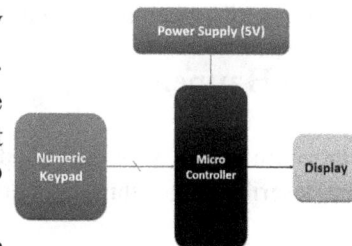

(iii)**Display :** The display used here is 3 Bi-colour LEDs. The glowingpatternofLEDsrepresentthedesiredminimizedexpresion. Fig. 56.1 : Power Supply

r1, r2, r3, r4, r5, r6, r7 r8, r9 = 3.3KΩ
r11=1KΩ
r10= 10KΩ variable

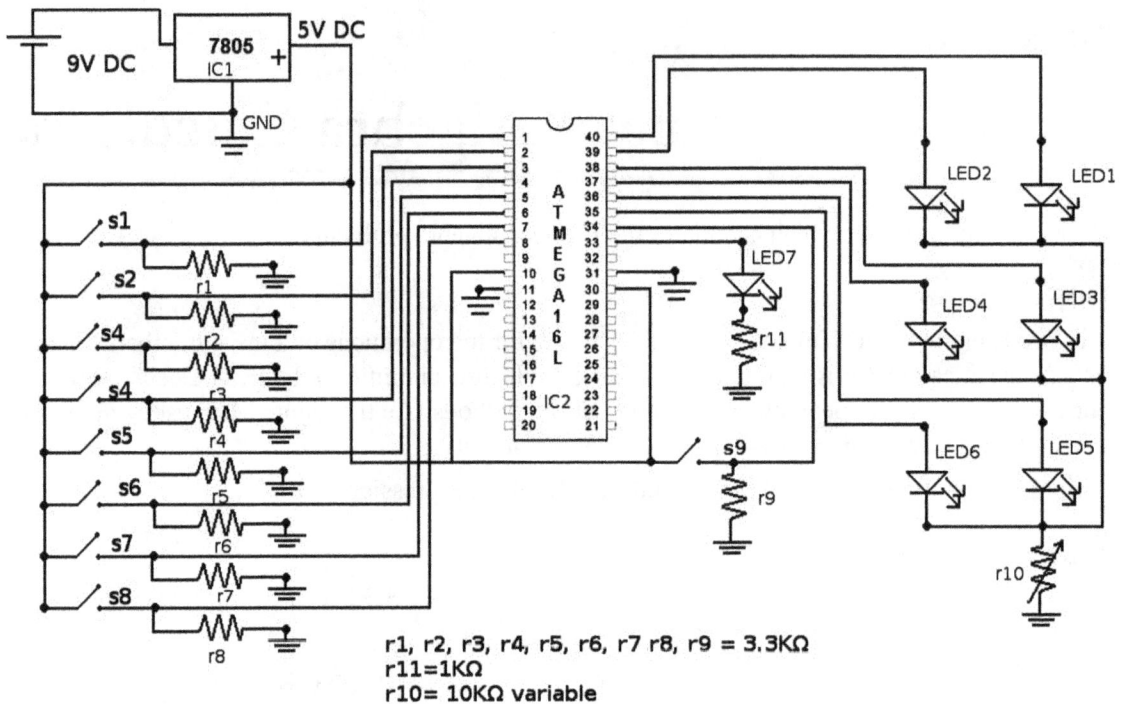

Keypad :
the input (minterms) expression.

To Do and Notice

The output is displayed as one minterm at a time. Next minterm is displayed by pressing the next

automatically when microcontroller is ready to take next input.

What Happens?

algorithm which was described. In this circuit micro-controller consist of code to implement the described algorithm and control other components in the circuit. Initially, when the power is

diagram.

Try It Yourself

Project-57

Car Parking Guard Circuit

How to design a circuit which helps the person in driving seat and give alarm if there is any obstacle comes in reverse ?

Introduction

While parking the car the driver should be more careful because he/she can not see the back of the car while parking or taking reverse. If there is any obstacle and ran over, it might cause damage to the car. The car parking guard circuit uses an infra-red sensor to guard in reverse direction.

Materials Required

R_1	:	10Ω, 2W carbon film resistor
R_2	:	2.2 kΩ, carbon film resistor
R_3	:	10 kΩ carbon film resistor
R_4	:	22 kΩ carbon film resistor
R_5	:	1 MΩ carbon film resistor
R_6	:	220 kΩ carbon film resistor
R_7	:	22 kΩ carbon film resistor
R_8	:	2.2 kΩ carbon film resistor
R_1, R_{10}	:	220 kΩ carbon film resistor
LED_1, LED_2	:	Infra-red emitting diode
LED	:	Light emitting diode
D_4, D_5	:	IN4001
D_6	:	IN4007
S_1	:	Lamp
PZ_1	:	Piezo buzzer
Q_1	:	Photo darlington transistor
Q_2	:	BD 140 PNP transistor
Q_3	:	BD 139 NPN transistor
C_1	:	0.1 μF capacitor
C_2	:	0.04 μF capacitor

6

7

8

1

Assembly

LM 567 : It is a tone detector which can interpret the frequency generated

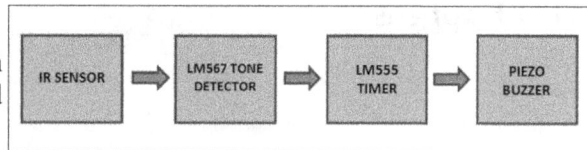

output according to the applications designed by the engineer. The tone dector is mainly used in touch tone decoders, ultrasonic controls, frequency montoring and control etc.

LM 555 :

etc.

IR sensor : The main function at IR sensor is to produce a beam for cerain distance. If there

Photo Darlington transistor

To Do and Notice

information to the tone detector.

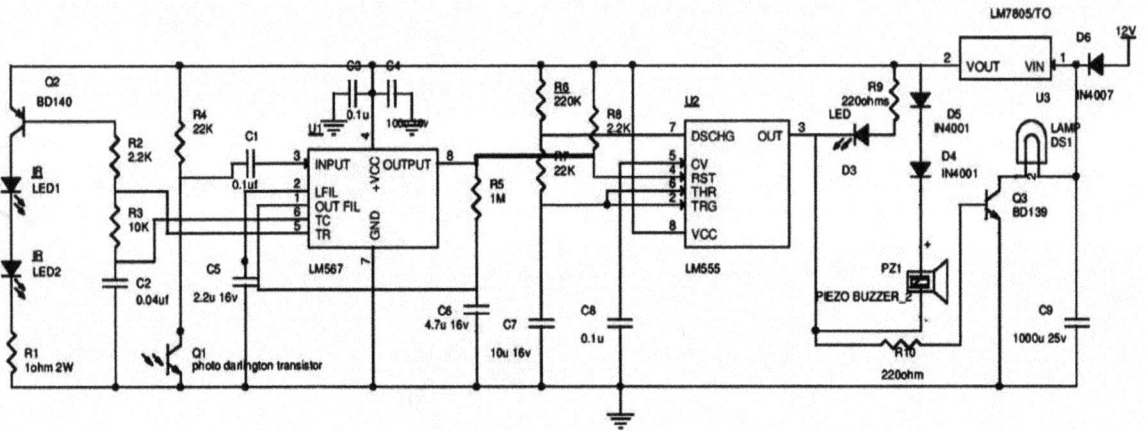

obstacle.

What Happens?

$_1$ will resit the IR sensor

$_6$, R_7 and capacitor

^{7} and resistor R

Try It Yourself
Make an automatic plant irrigation system in which there is no need of human interference.

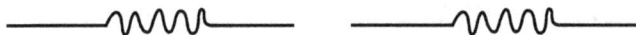

Project-58

Audio Power Amplifier

How to amplify an audio signal ?

Introduction

The audio power amplifier project is based on LM 1875 amplifier module. It is able to deliver up to 30W of power using an 8Ω load and dual 30V DC power supplies. It is design to operate with minimum external components with current limit and thermal shutdown protection features. Other features include high gain, fast slow rate, wide power supply range, large output voltage swing and high current capability.

Materials Required

U_1	:	LM 875 audio power amplifier
SPK_1	:	8 Ω speaker
T_1	:	240V/36V 80A centre topped transformer
D_1, D_2, D_3, D_4	:	1N5401 diode
V_1	:	275V diameter 14 mm varistor
S_1	:	Power switch
F_1	:	5A 240V AC fuse
F_2	:	2A 240V AC fuse
E_1, E_2, E_3, E_4	:	2200 µF/63V electrolytic capacitor
E_5	:	1 µF/63V electrolytic capacitor
E_6	:	22 µF/63V electrolytic capacitor
C_1, C_2, C_3, C_4	:	0.1 µF/50V ceramic capacitor
C_5	:	0.22 µF Polyester capacitor
R_1	:	1 µF 1/4W carbon film resistor
R_2	:	1 MΩ 1/4W carbon film resistor
R_3	:	22 kΩ 1/4W carbon film resistor
R_4	:	20 kΩ 1/4W carbon film resistor
R_5	:	1 kΩ 1/4W carbon film resistor
R_6	:	1 kΩ 1/4W carbon film resistor
Heat Sink	:	Termal resistance of 1/4" C/W or above screw, nut, washer to mount heat sink to LM1875 insulated TO-220 washer

Assembly

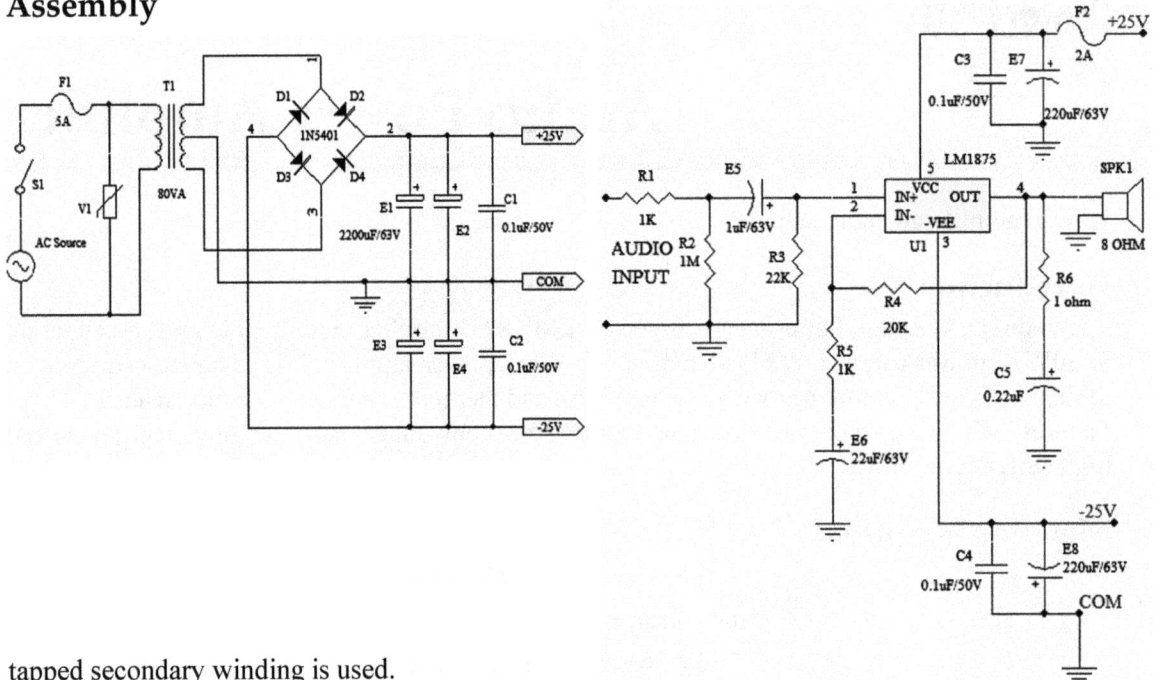

tapped secondary winding is used.

protect the circuit against power surge.

To Do and Notice

signal.

What Happens?

$_1$ and

electrolytic capacitor E
loudspeaker. Resister R_6

module will be cut-off from operation due to the heat that will build up during the operation of the

power supply; hence, it must be isolated from the heat sink by the use of an insulating washer. If

Try It Yourself

Project-59

Touch Switch

How to make a LED ON or OFF by a physical touch ?

Introduction

In this project a CMOS quad 2 input NOR gate IC is used as a latching circuit to switch a LED ON and OFF by physically touching the ON metal plate or OFF metal plate. The CD4001BC integrated circuit is a monolithic complementary MS (CMOS) IC that are constructed with N- and P- channel enhancement mode transistors. Its input are protected against electrostatic discharge with diodes to VDD and VSS.

Materials Required

U_1	:	Quad 2 input NOR gate CMOS CD4001BC
LED	:	Light emitting diode
R_1	:	3.3M Ω 1/4W, 5% resistor
R_2	:	3.3M Ω 1/4W, 5% resistor
R_3	:	10 kΩ 1/4W, 5% resistor
R_4	:	470 kΩ 1/4W, 5% resistor
R_5	:	10 kΩ 1/4W, 5% resistor
C_1	:	0.1 μF/25V ceramic capacitor
BAT	:	9V battery with holder
T_1, T_2, T_3, T_4	:	Touch terminal
Q_1	:	PNP transistor 2SA953

Assembly

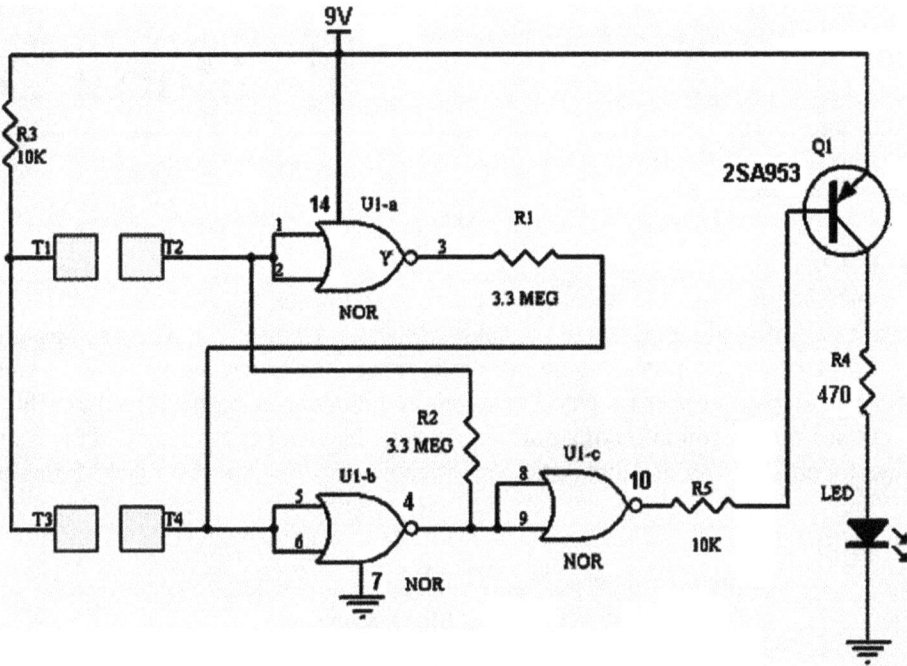

To Do and Notice

$_1$ and T or T and T

What Happens?

$_1$ and T

and T are bridged of which the
$_1$-c will go to

Try It Yourself

Project-60

Morse Code

Do you know how a morse code send and receive ?

Introduction

The Morse code is a method of transmitting text information as a series of on-off tones, lights, or clicks that can be directly understood by a skilled listener or observer. The international morse encodes the ISO basic latin alphabet some extra latin letters, the arabic numerals and a small set of punctuation. Each character is represented by a unique sequence of dots and dashes. The duration of a dash is three times the duration of a dot. This is done by using 555 timer IC. This is one of the various ways for amature radio enthusiasts to practice the sending and receiving of morse code.

Material Required

U_1	:	NE 555 timer
R_1	:	10 kΩ 1/4W, 5% carbon film resistor
R_2	:	47 kΩ 1/4W, 5% carbon film resistor
VR_1	:	100 kΩ potentiometer
VR_2	:	10 kΩ potentiometer
C_1	:	0.01 μF/25V ceramic capacitor
E_1	:	33 μF/25V electrolytic capacitor
SPKR	:	8 Ω speaker
S_1	:	Switch
KEY	:	Key
BAT	:	9V Battery, Battery holder

Assembly

In this schematic, a 555 IC is used and configured as a timer astable mode. Once triggered, it will generate a frequency from its output at Pin 3. Once, a key is pressed, it will drive the 8Ω loudspeaker which is connected in parallel to potentiometer VR_2.

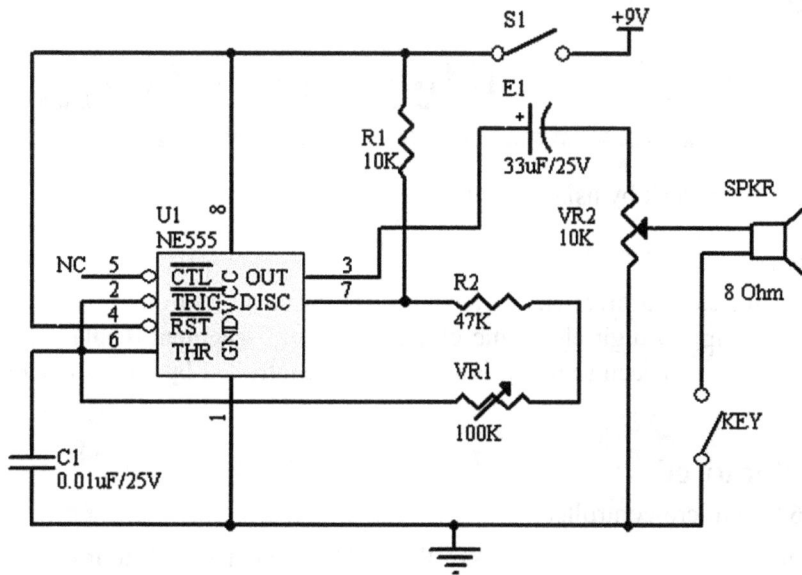

To Do and Notice

dit or *dah* sound when the key

is pressed.

What Happens?

1

1

1

1. The
which is connected is parallel

with speaker and the key.

Try It Yourself
Try to make a Ham Radio.

Project-61

Electronic Timer Clock

Can you make a time clock by using PIC6C54 micro-controller ?

Introduction

This electronic time clock project uses a 18 pin PIC16C54 microcontroller as its core in the display and setting of a simple 4 digit electronic clock. The four 7 segment displays are connected in multiplexing method and each segment is sequentially controlled by common cathode of each 7 segments.

Materials Required

18 Pin PIC 16 C54 micro-controller

Resistors	:	Four 10 kΩ, three 820 Ω, eight 100 Ω, four 4 kΩ
Transistors	:	4 PNP 2N5401 transistors
Display	:	4 seven segment display LEDs
Switch	:	3 switch SW_1, SW_2, SW_3, 1 relay switch
Battery	:	5V DC battery

Assembly

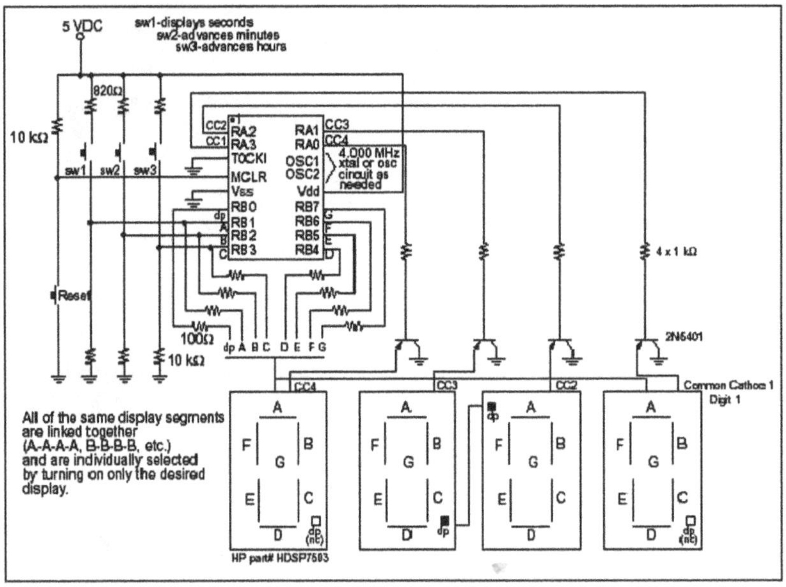

Fig. 61.1 : Circuit Diagram

To Do and Notice

What Happens?

Multiplexing helps to reduce the number of parts needed to connect to the 7 segments. It also keeps

Try It Yourself

Project-62

Temperature Switch

Do you have understanding of the use of germanium diode and silicon diode ?

Introduction

This temperature switch project will provide you with an understanding of how to use of germanium diode and how it works compared to the more common silicon diode. It works on the principle that as the temperature surrounding the germanium diode increases. The back resistance decreases sharply.

Materials Required

Q_1, Q_2	:	BC 548 NPN transistor
Q_3	:	2 SC2002 NPN transistor
R_1	:	120 kΩ, 1/4W 5% carbon film resistor
R_2	:	1 kΩ, 1/4W 5% carbon film resistor
VR_1	:	1 MΩ potentiometer
D_1	:	Germanium diode
D_2, D_3	:	1N4003 diode
BZ	:	12 Buzzer

Assembly

The schematic of the project is shown in fig 62.1. It uses only 10 components to illustrate the understanding of germanium diode and its characteristics. This project is suitable for beginners to electronics and will definitely help to inculcate the interest in designing electronic projects.

You can use any equivalent NPN and PNP transistors to replace BC548 and 2SC 2002 respectively as these parts may not be freely available in our country.

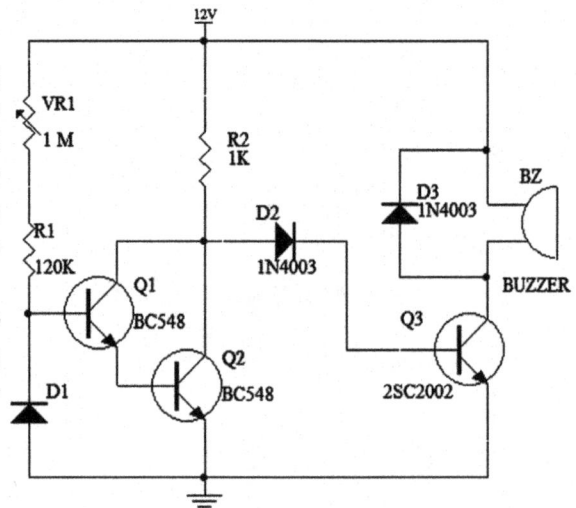

Fig. 62.1 : Circuit Diagram

To Do and Notice

What Happens?

1

1

1 to pull down near ground potential. This cause[1]
is now forward bias through resistor R and

1   1

Try It Yourself

diodes.

Project-63

Phone in Use

How will you determine the status of a phone line traffic by glowing two LEDs of different colours?

Introduction

This project is known as phone in use which is a simple indicator that you can design and construct that displays the status of the phone line. If the line is in use, the yellow LED will turn ON. If it is not in use, the green LED will turn ON.

By having this indicator the user will not be interrupted by another user who wants to use the same line.

Materials Required

R_1	:	3.3 kΩ, 1/4W, 5% resistor
R_2	:	18 kΩ, 1/4W, 5% resistor
R_3	:	22 kΩ, 1/4W, 5% resistor
R_4	:	68 kΩ, 1/4W, 5% resistor
R_5	:	5.6 kΩ, 1/4W, 5% resistor
Q_1, Q_2	:	2N3393 NPN transistor
L_1	:	3 mm LED, yellow colour
L_2	:	3 mm LED, green colour
D_1, D_2, D_3, D_4	:	1N4002 diode

Assembly

As shown in the circuit, dioes D_1, D_2, D_3 and D_4 are used to ensure that in the event that the tip and ring of the line is reversed, the circuit can still be used.

When the telephone is in on hook condition, the voltage across the tip and ring is in the range of 48V DC to 50V. When it is in off-hook condition (the receiver is taken off its hook), the voltage drops to the range of 6V DC to 15V DC.

Fig 63.1 :

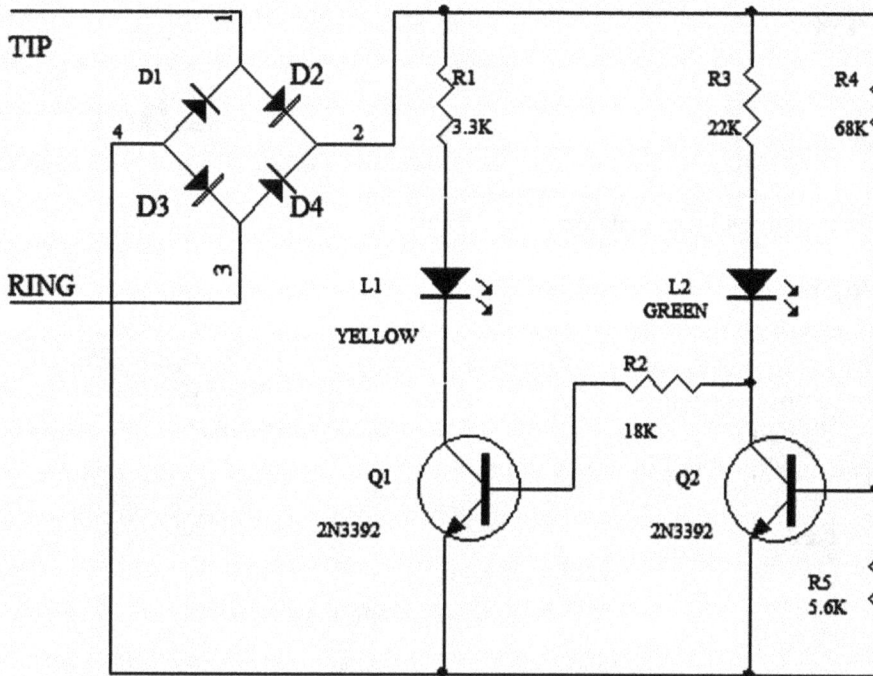

To Do and Notice

[1] will turn

What Happens?

and R

[1]

Try It Yourself

Project-64

Thermostat

How to make an automatic room temperature control system?

Introduction

This project focuses on the heating control of a space that uses electric heater as its source of heating. It basically consists of a comparator that controls the ON and OFF of the electric heater based on the sensor temperature. A thermostat is used to control and regulate the temperature of a space to its desired set point. They are used entensively in HVACR which stands for heating, ventilation, air conditioning and refrigeration.

Materials Required

RLY	:	SPDT 12V relay
U_1	:	LM 358 operational amplifier
R_1	:	43 kΩ 1/4W, 1% resistor
R_2	:	24 kΩ 1/4W, 1% resistor
R_3	:	22 kΩ 1/4W, 1% resistor
R_4	:	22 kΩ 1/4W, 1% resistor
R_5	:	3.9 mΩ 1/4W, 5% resistor
R_6	:	5.1 kΩ 1/4W, 5% resistor
R_7	:	33 kΩ 1/4W, 1% resistor
VR_1	:	5 kΩ potentiometer
VR_2	:	5 kΩ slide or rotary liner potentiometer
ZD_1	:	Zener diode 5.6V 0.25 W
Q_1	:	NPN transistor 2SC 2002
Thermistor	:	Termister 20 kΩ @ 25°C, B > 4000kW
HTR	:	Heater

Assembly

To Do and Notice

off and heater will conduct until the temperature of the room rises again.

What Happened?

resistance will go up and hence the output of the

the temperature of the room rises again.

potentiometer VR

1

$_1$, set the resistance

to the load of the heater.

Try It Yourself

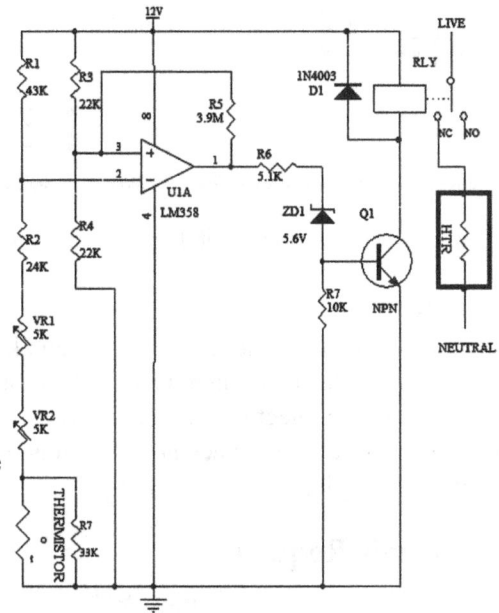

Project-65

Electronic Timer Switch

Can you make a timer switch which turns a lamp ON and OFF periodically ?

Introduction

A timer switch is a timer that operates an electric switch controlled by the timing mechanism. The timer may switch equipment ON, OFF, or both at a preset time after a preset interval or cyclically. This is a good project to simulate the presence of occupants in a house. These days when security is becoming a concern when no one is at home, having this device at home will deter the thief away from breaking in.

Materials Required

U_1	:	CD4060B binary counter
R_1	:	100 kΩ 1/4W, 5% carbon film resistor
R_2	:	2.2kΩ 1/4W, 1% carbon film resistor
R_3	:	1 M kΩ 1/4W, 1% carbon film resistor
R_4	:	3.9kΩ 1/4W, 5% carbon film resistor
R_5	:	10 kΩ 1/4W, 5% carbon film resistor
D_1, D_2, D_3, D_4	:	1N4148 diode
D_5	:	1N4003 diode
C_1	:	0.01 μF/25V ceramic capacitor
C_2	:	0.1 μF/25V ceramic capacitor
K_1	:	SPDT 12V relay with 5A/240 V AC contact rating
LAMP	:	Lamp

Assembly

The core of this electronic timer switch project uses a CD4060B binary counter. The binary counter has 10 outputs and the counter is counted by configuring the oscillator. Every negative clock will trigger the counter of the IC internally.

The timing of the circuit is affected by resistor R_3 and capacitor C_2. By connecting the four outputs in an AND configuration, the transistor Q_1 will only turn ON if all the 4 outputs are in logic '1'. If any of the logic is '0' the transistor will remain OFF.

Fig :

Electronic Timer Switch

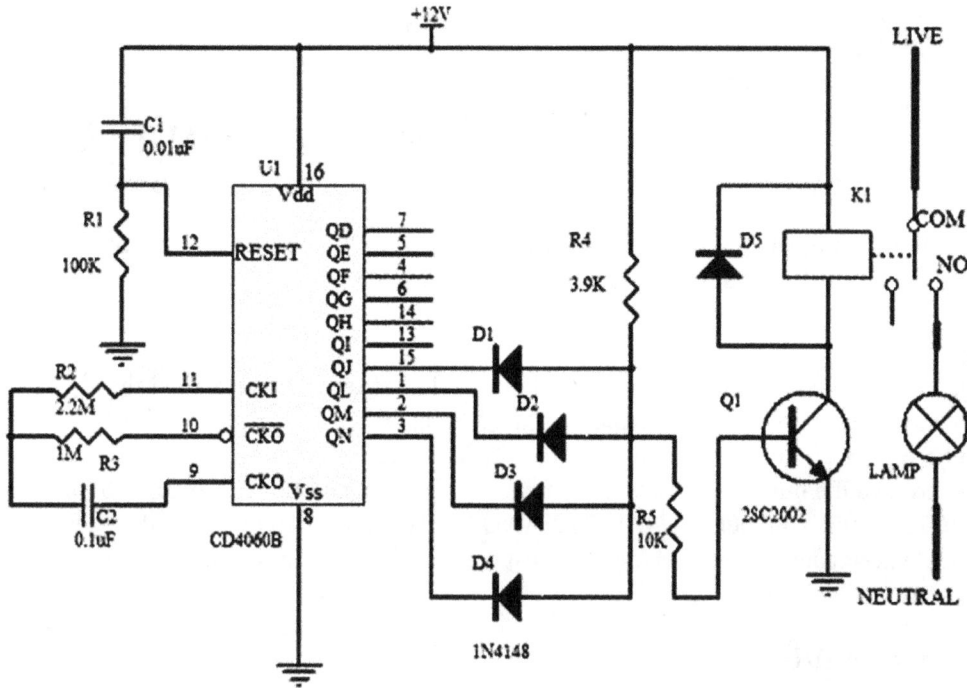

To Do and Notice

What Happens?

$_1$ will switch status accordingly. The timing of the switching can be changed by changing , R .

Note that since the oscillator is not using crystal, the timing may not be as accurate compared to

Try It Yourself
Extend the

Project-66

Educational Game

Can you make a project which is used in a quiz competition to see which of the participants press the button first to answer questions posed by the quiz master ?

Introduction

Quiz games are becoming very popular these days. The main concept behind these games is the use of fastest finger first indicator which are used to test the reaction time of the contestants. The project here has a circuit that can be used for 2 players. It can be used to identify which of the two has first pressed the button and locks out the other. This constructional project is based on a latch CD74HC75 IC and a 4 input CD74AC 20 NAND IC. The first person to press the switch will light up the LED designated for the person. The rest of theLEDs will not be activated until the RESET button is pressed.

Materials Required

U_1	:	CD74HC 75 or equivalent
U_2	:	CD74AC 20 or equivalent
R_1, R_2, R_3, R_4, R_5	:	1.5 kΩ 1/4W, 5% carbon film resistor
R_6, R_7, R_8, R_9	:	470 kΩ 1/4W, 5% carbon film resistor
$LED_1, LED_2, LED_3, LED_4$	:	5 mm or 3 mm LED
RESET, S_1, S_2, S_3, S_4	:	Normally open push button
C_1	:	0.1 μF/25V ceramic capacitor

Assembly

The truth table of the latch IC and the NAND IC is as shown in the figures below.

Inputs		Outputs	
D	E	Q	$\overline{Q}$
L	H	L	H
H	H	H	L
X	L	Q_0	$\overline{Q}_0$

Fig. 66.1 : CD74HC 75 Latch Function Table

Inputs				Outputs
				Y
H	H	H	H	
	X	X	X	H
X		X	X	H
X	X		X	H
X	X	X		H

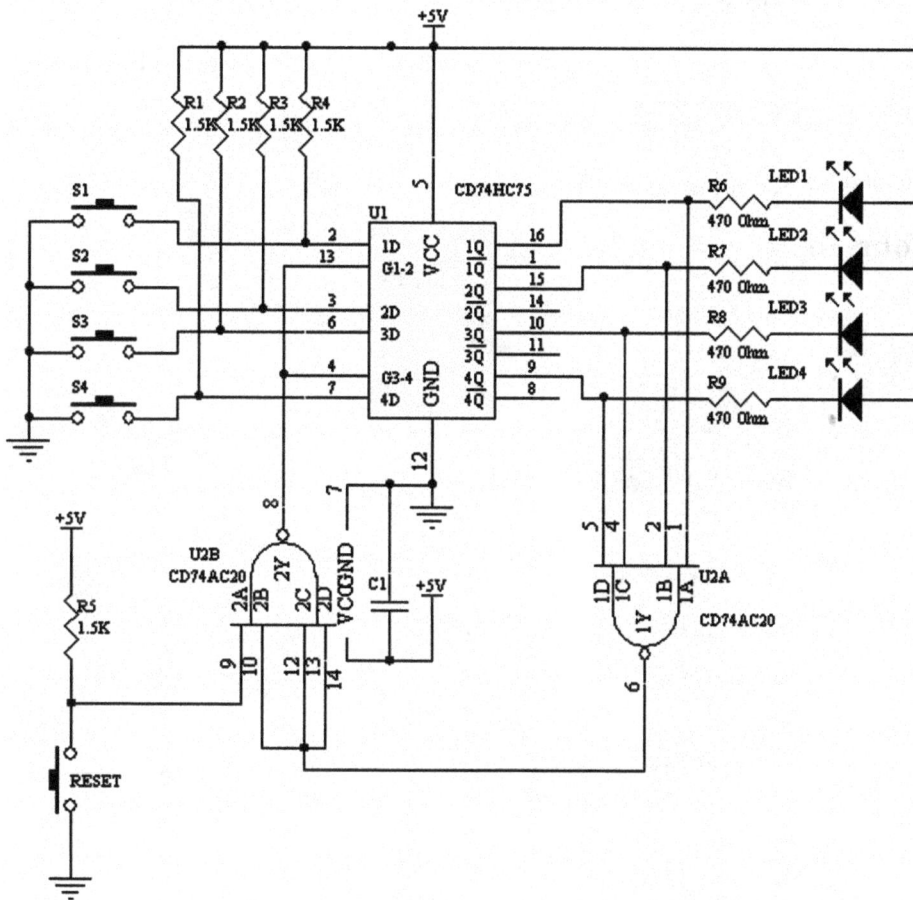

+5V

R1 1.5K R2 1.5K R3 1.5K R4 1.5K

S1 S2 S3 S4

U1 CD74HC75

1D G1-2 VCC
2D 3D
G3-4 4D GND

2 13 3 6 4 7 5 12

1Q 1Q 2Q 2Q 3Q 3Q 4Q 4Q
16 1 15 14 10 11 9 8

R6 470 Ohm LED1
R7 470 Ohm LED2
R8 470 Ohm LED3
R9 470 Ohm LED4

U2B CD74AC20
2A 2B 2C 2D 2Y
VCC GND
C1 +5V
9 10 12 13 14 8 7 12

U2A CD74AC20
1D 1C 1B 1A 1Y
5 4 2 1 6

+5V
R5 1.5K
RESET

To Do and Notice

$_1$ is a result of the player who pressed S_1, it will be shown as the one who

What Happens?

output any data once any of the button is pushed.

$_1$ $_1$ $_1$ to light

$_1$

to press the button.

Try It Yourself

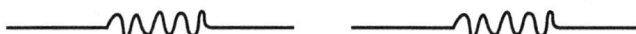

Project-67

Auto Shut-off Tone Generator

Can you build an emergency siren which is shut off until the duration of the set has elapsed?

Introduction

In this auto shut-off tone generator project, once the switch to the 9V power supply is connected, the alarm will trigger at a frequency of approximately 1.27 kHz.

It will remain ON for a duration of approximately 2.8 minutes before it stopped. This project is useful when built as one can carry it along wherever one goes or places it in a vehicle. In times of emergency, one can easily switch ON the switch and the loud speaker will emit a loud sound that will frighten the uninvited guest.

Materials Required

U_1, U_2	:	NE 555 timer
R_1	:	18 kΩ 1/4W, 5% carbon film resistor
R_2	:	330 kΩ 1/4W, 5% carbon film resistor
R_3	:	47 kΩ 1/4W, 5% carbon film resistor
R_4	:	33 kΩ 1/4W, 5% carbon film resistor
R_5	:	220 kΩ 1/4W, 5% carbon film resistor
E_1	:	4.7 µF/25 V electrolytic capacitor
E_2	:	470 µF/25 V electrolytic capacitor
C_1	:	10 nF/25V ceramic capacitor
Q_1	:	TIP41A
SPKR	:	8Ω speaker
S_1	:	switch
BAT	:	9 V battery, battery holder

Assembly

The schematic of the tone generator is as shown the fig 67.1. It is based on two 555 timer ICs or one single 556 timer IC (which contains two 555 timers). In this schematic, two 555 timers are used. V2 is configured as a timer in astable mode.

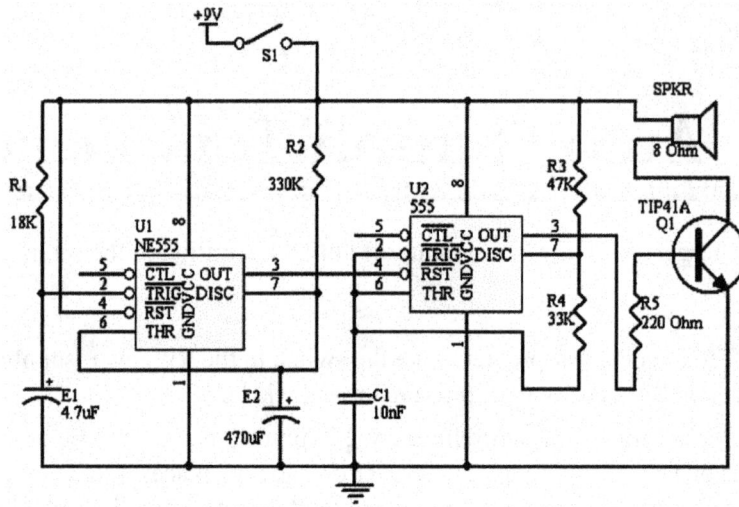

To Do and Notice

. Hence the loud sound
emitted by the 8 loud speaker stops emitting sound.

What Happens?

$_1$ $_1$

8 loud speaker to emit a loud audible sound.

and

mode, the frequency of the sound can be obtained.

$_1$

$_1$

$_1$ is immediately triggered.

Try It Yourself

Project-68

Analog Delay Timer

Can you build a project for devices i.e. compressors and halogen lamps which can not be OFF and ON repeatedly within a short period of time as it will cause damage to the devices ?

Introduction
In this delay timer project, all analog parts are being used with the thyristor as a device that switches an AC relay ON or OFF depending on the timing of the RC circuit. The input mains supply used ranges from 220V AC to 240 V AC and an AC relay (220-240V AC) is used to switch a load. The load to be switched must be within the current and relay ratings. This project is useful for the use of devices that need to be OFF for a minimum of 150-120 seconds after the mains supply has cut-off.

Materials Required

SCR	:	Thyristor 2P6M,MCR 106-8 or equivalent
$D_1 - D_9$	:	1N4001 diodes
$Q_1 - Q_2$	:	2SC2002 NPN transistor
Q_3	:	2SA953 PNP transistor
E_1	:	10 μF electrolytic capacitor
E_2	:	47 μF electrolytic capacitor
C_1	:	0.47 μF ceramic disc capacitor
RLY	:	AC relay
Z	:	9.1V zener diode
R_1	:	47 kΩ resistor
R_2	:	4.7 kΩ resistor
R_3	:	2.2 kΩ resistor
R_4	:	100 kΩ resistor
R_5	:	100 kΩ resistor
R_6	:	100 kΩ resistor
R_7	:	MΩ resistor
R_8	:	47 kΩ resistor
R_9	:	100 kΩ resistor
R_{10}	:	4.7 kΩ resistor

R_{11} resistor

Assembly

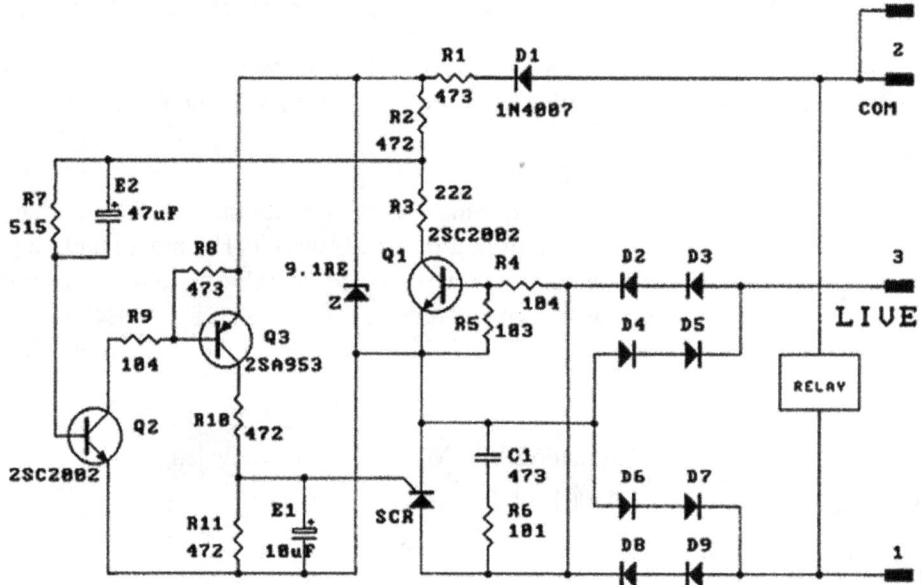

7 and electrotylic capacitor E . More

To Do and Notice

has been discharged through resistor R_7.

7 and E .

What Happens?

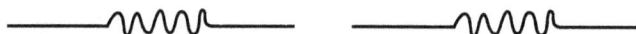

$_1$ and R$_6$ across the

7

Try It Yourself

Project-69

Room Thermostat

Can you make a system which controls the ON and OFF of the chilled water valve based on the sensor temperature ?

Introduction

A thermostat is a component of a control system which senses the temperature of a system so that the temperature of system is maintained near a desired set point. The thermostat does this by switching heating or cooling devices ON or OFF, or regulating the flow of a heat transfer fluid as needed to maintain the correct temperature. Thermostats can be constructed in many ways and may use a variety of sensors to measure the temperature. This room thermostat project will focus on the chilled water control of a space that uses chilled water as its source of cooling. The controller basically consists of a comparator that controls the ON/OFF of the chilled water valve based on the sensor temperature.

Materials Required

RLY	:	SPDT 12V relay
U_1	:	LM358 operational amplifier
R_1	:	43 kΩ, 1/4W, 1% resistor
R_2	:	24 kΩ, 1/4W, 1% resistor
R_3	:	22 kΩ, 1/4W, 1% resistor
R_4	:	22 kΩ, 1/4W, 1% resistor
R_5	:	3.9 kΩ, 1/4W, 5% resistor
R_6	:	5.1 kΩ, 1/4W, 5% resistor
R_7	:	33 kΩ, 1/4W, 1% resistor
VR_1	:	5 kΩ potentiometer
VR_2	:	5 kΩ slide or rotary linear potentiometer
VD_1	:	Zener diode 5.6 V 0.25 W
Q_1	:	NPN transistor 2SC2002
Thermistor	:	Thermistor 20 kΩ @25°C, B > 74000 k
CW valve	:	Chilled water valve

Assembly

a comparator to sense the inputs of the reference

To Do and Notice

pass through the coil.

The cold air from the coil is transferred to the fan until the temperature of the room goes down to its set point again.

What Happens?

turn

Try It Yourself

Project-70

Time Delay Circuit

Can you make a simple time delay circuit ?

Introduction

In the design of analog circuits, there are times when you will need to delay a pulse that comes into a circuit before being used for the next process. This circuit uses a 555 timer to delay a pulse that comes into a maximum time of 75 seconds.

The timing of delay can also be changed by changing the resistor value of VR_1 and the capacitor value of E based on the time delay formula of $t = 0.69\ RC$.

Materials Required

U_1	:	555 timer
R_1	:	470 kΩ 1/4 W 5% carbon film resistor
VR_1	:	5 MΩ variable resistor
Q_1	:	BC 556 PNP transistor
E	:	22 μF/25V electrolytic capacitor

Assembly

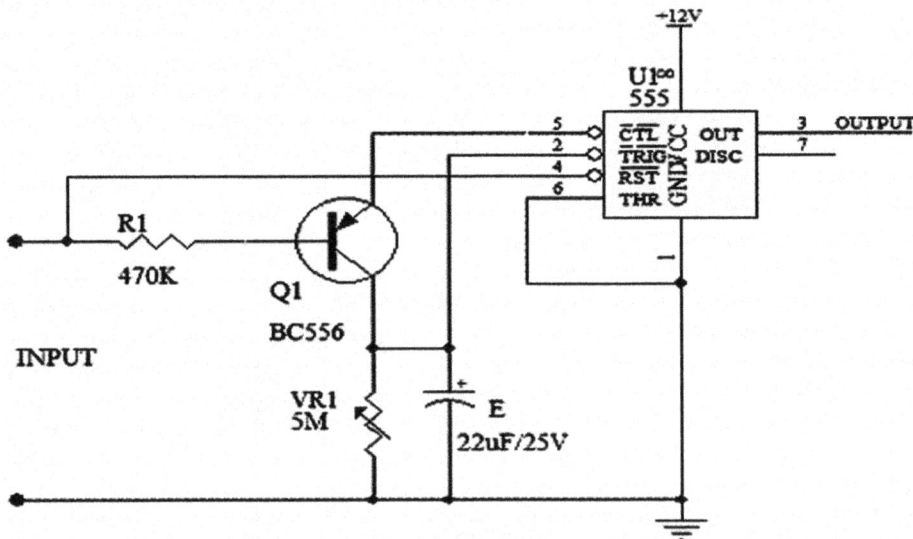

Fig. 70.1 : Circuit Diagram

To Do and Notice

₁ will be discharged through VR_1.

₁

What Happens?

₁ will turn

This causes the capacitor E_1 to be discharged through VR_1 resistor. The time delay will depend on

₁ and E to get different time delay.

seconds.

Try It Yourself

Project-71

Cooler Pump Protection cum Humidity Control

Can you protect your cooler pump and control humidity ?

Introduction

The desert cooler pumps pack off mainly due to exessive heating during summer peaks. Another short coming of these coolers is excessive humidity. Yet another problem faced particularly with low water capacity coolers is that you may suddenly discover at past mid night that there is no water in the cooler tank. All these problems could be overcome if the cooler pump is operated in a duty cycle while the fan was run continuously. Typically, the pump may be kept ON for two minues and then switched OFF for a time ranging from four to eight minutes depending upon environmental temperature. This not only enhances pump life but also controls humidity. In fact, with humidity control, cooling effect is better.

Materials Required

R_1, R_2	:	1 MΩ, 1/4 W carbon film resistor
R_3	:	2.2 kΩ, 1/4 W carbon film resistor
P_1	:	2.2 kΩ, potentiometer (LIN)
C_1	:	220 μF, 25V electrolytic capacitor
C_2	:	0.01 μF, 25V ceramic disc capacitor
C_3	:	1000 μF, 25V electrolytic capacitor
C_4	:	0.1 μF ceramic disc capacitor
D_1, D_2, D_3	:	IN4001 or equivalent diodes
VR-1	:	Three terminal regulator, type 7812
IC-1	:	555 timer
Relay	:	12V DC relay with at least one normally open contact
T-1	:	12V, 250 mA mains transformer
Fuse	:	0.25 A rating
S-1	:	Simple ON/OFF toggle switch

Assembly

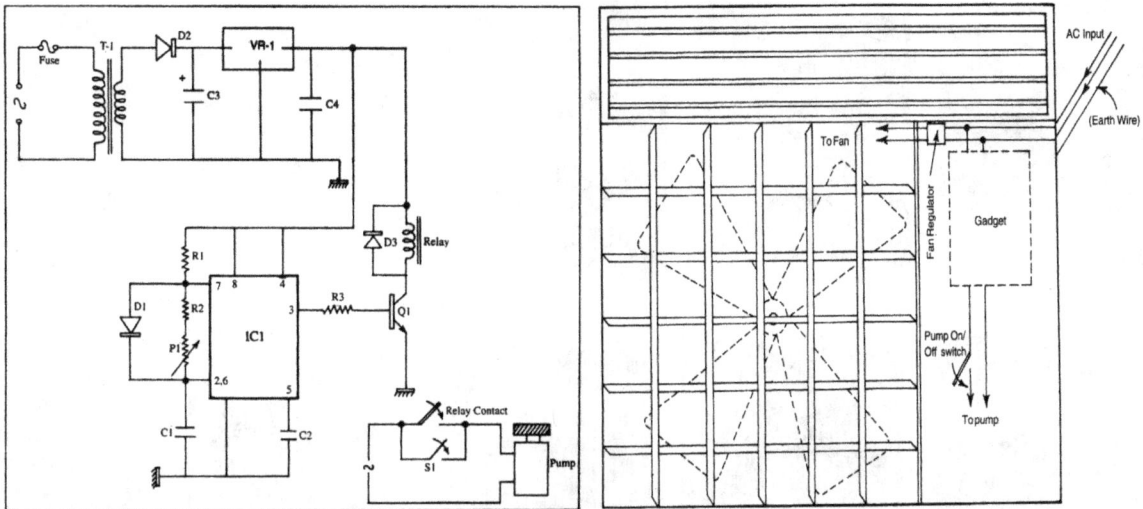

going to the cooler fan through the regulator) also goes to the input of the gadget. The assembled unit should be mounted on the inside of the front panel near the usual controls of the cooler. The

not get inside the unit.

To Do and Notice

Switch S-1 acrosss the relay contact can be used in case you desire to run water pump continuously.

What Happens?

1   1   1   1

S–1 is used to bypass the relay contact in case you want to run the pump continuously.

Try It Yourself

Make a gadget which cuts-off the water pump when the water tank goes empty.

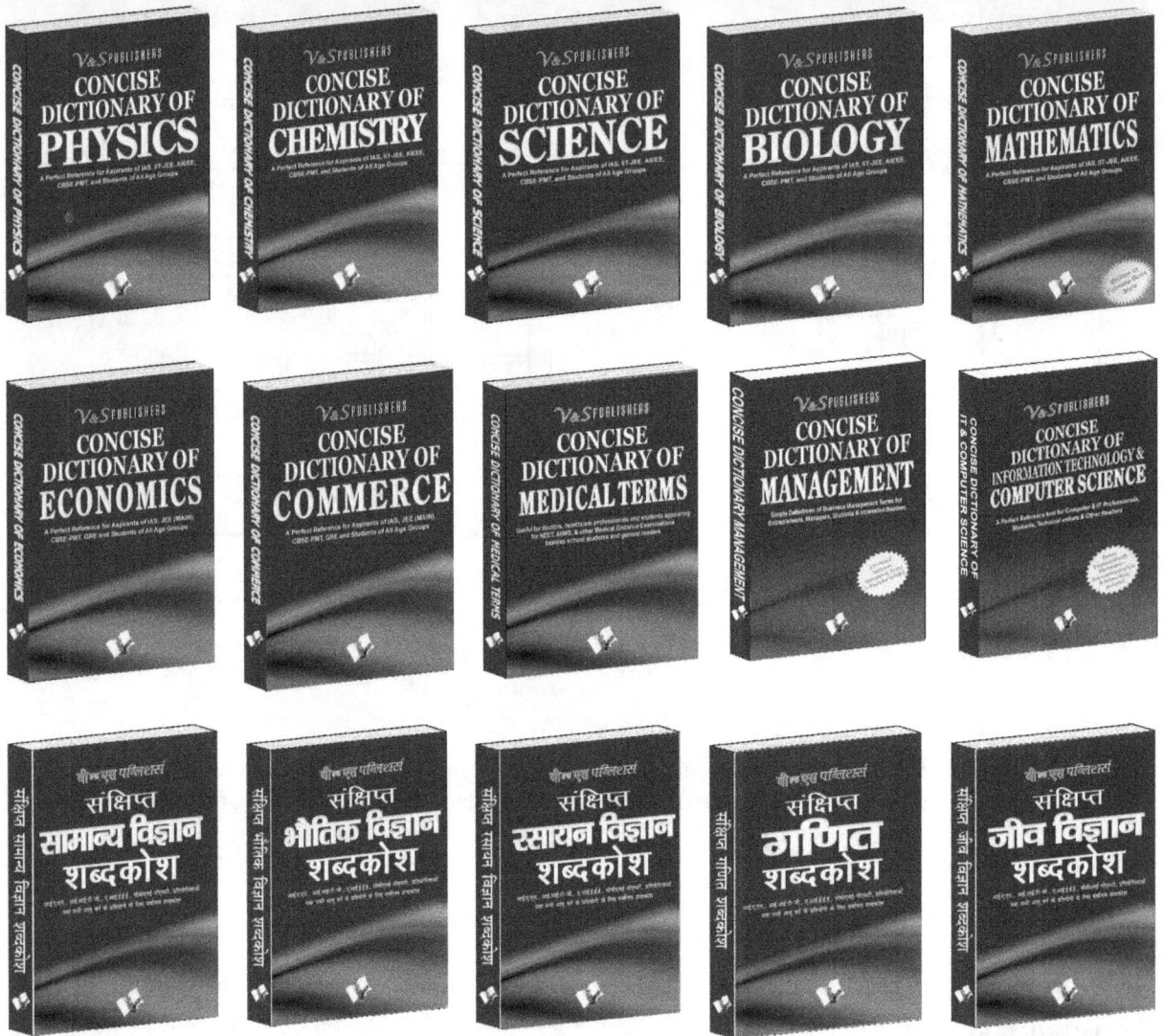

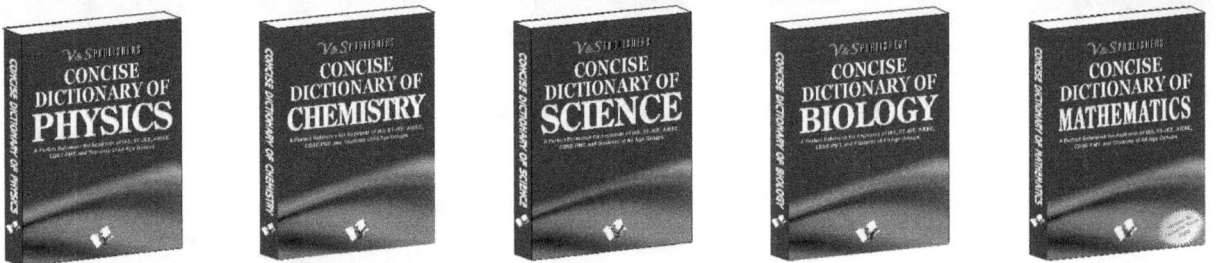

CONCISE DICTIONARIES
(संक्षिप्त शब्दकोश)

English-English Dictionaries

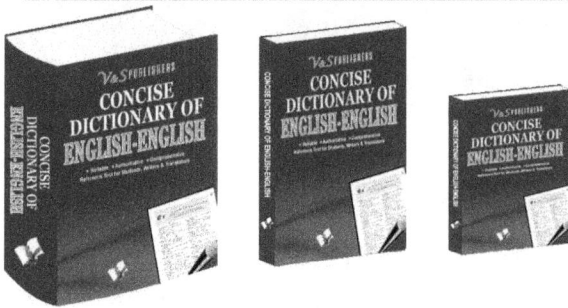

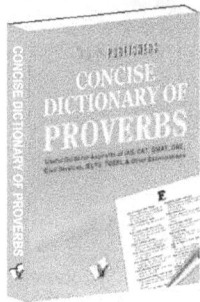

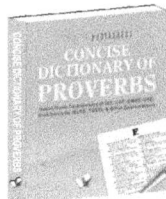

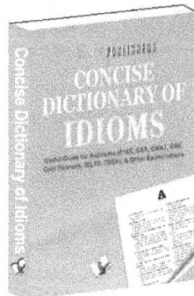

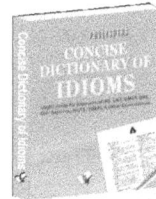

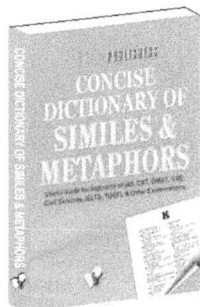

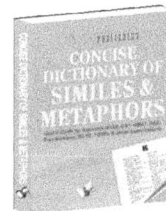

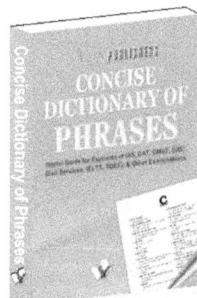

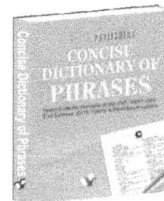

English-English-Hindi Dictionaries

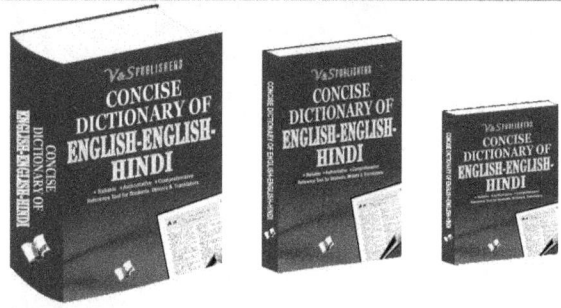

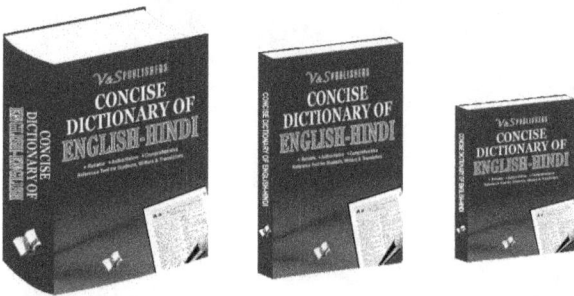

English-Hindi Dictionaries

Hindi-English Dictionaries

ENGLISH DICTIONARIES

www.ingramcontent.com/pod-product-compliance
Lightning Source LLC
Chambersburg PA
CBHW080532220326
41599CB00032B/6289